Martin Fiedler

Tesis doctoral en el Programa Oficial de Doctorado en
Procesos y Productos Químicos

Tribological and rheological investigation of the applicability of biogenic lubricating greases as separating medium in highly stressed friction systems

Supervisor 1:
Dr. José María Franco Gómez

Supervisor 2:
Dr. Erik Kuhn

Date of defense: November 6, 2014

University of Huelva, Department of Chemical Engineering, Physical Chemistry and Organic Chemistry

Tribological and rheological investigation of the applicability of biogenic lubricating greases as separating medium in highly stressed friction systems

Report submitted by Martin Fiedler to qualify for the Degree of Doctor with international mention at the University of Huelva

M.Eng. Dipl.-Ing.(FH) Martin Fiedler[*]

October 6, 2014

This thesis has been performed at the Department of Chemical Engineering, Physical Chemistry and Organic Chemistry at the University of Huelva under the direction of Dr. José María Franco Gómez and Dr. Erik Kuhn, who authorize filing

---

[*]Hochschule für Angewandte Wissenschaften Hamburg, Fakultät Technik und Informatik, Department Maschinenbau und Produktion, Berliner Tor 5, 20099 Hamburg, Germany

Resumen

Actualmente existe una gran demanda de grasas lubricantes medioambientalmente amigables que sólo puede ser satisfecha parcialmente. El propósito de este trabajo es investigar, fundamentalmente, el efecto que los principales componentes de diferentes bio-grasas lubricantes ejercen sobre las características reológicas y tribológicas resultantes. Se ha seleccionado un número limitado de componentes, cuatro aceites base diferentes y cuatro espesantes con diferente estructura, que califican a estas grasas como bio-lubricantes. Las dieciséis grasas analizadas han sido formuladas con idéntico grado de NLGI con el fin de eliminar la influencia de la consistencia en los resultados. Se ha estudiado, en particular, el efecto de la polaridad de los diferentes aceites base y su interacción con los diferentes espesantes.

Las grasas han sido sometidas a experimentos tribológicos en un nano-tribómetro aplicando cuatro fuerzas normales diferentes y dos combinaciones diferentes de materiales que dan como resultado un conjunto de esfuerzos hertzianos para cada grasa estudiada.

Adicionalmente, se ha llevado a cabo un análisis de microscopía de fuerza atómica con el fin de clarificar la microestructura de cada una de ellas.

Asimismo, las grasas han sido reológicamente caracterizadas aplicando diferentes situaciones de deformación. Por un lado, el comportamiento de flujo en modo transitorio fue monitorizado en ensayos rotacionales y, por otro lado, las grasas fueron reológicamente caracterizadas a través de ensayos oscilatorios de baja amplitud en función de la frecuencia y la amplitud aplicadas. Estos resultados fueron también evaluados desde un punto de vista energético.

Además, se ha realizado una serie de experimentos no convencionales tales como los ensayos de tensión extensional, realizados en un reómetro usando una geometría placa-placa donde se fuerza a la separación de las placas registrando la fuerza normal aplicada, y experimentos de rebote donde una bola de acero golpea una superficie también de acero, lubricada con grasa, después de caer desde una distancia determinada. En estos últimos ensayos se ha evaluado la altura de rebote y la marca del impacto.

Los principales hallazgos de este trabajo se refieren a la influencia que ejerce la polaridad de los diferentes aceites sobre las características tribológicas y reológicas de las grasas.

En primer lugar, los experimentos tribológicos muestran el tipo, la intensidad y la extensión de las marcas de desgaste producidas entre las dos superficies en contacto. Se ha demostrado que la influencia de la polaridad del aceite base depende en gran manera del tipo de espesante y de la combinación de materiales utilizados. También se demuestra que la influencia de la polaridad puede ser reducida dependiendo de la combinación de materiales.

En segundo lugar, la influencia de la polaridad fue detectada en los resultados reológicos de las grasas. Se ha demostrado igualmente que el efecto de la polaridad está altamente influenciado por el espesante utilizado. Por otra parte, la polaridad del aceite afecta el porcentaje de espesante necesario para alcanzar una determinada consistencia, dependiendo de la naturaleza de éste. En los ensayos de flujo transitorio se ha intentado eliminar la dependencia del porcentaje de espesante. Esto se ha realizado mediante la evaluación de la energía de activación necesaria para la degradación estructural del espesante, mostrando una gran diferencia entre grasas formuladas con aceites altamente polares respecto a aquellas basadas en aceites con baja polaridad. Nuevamente, se ha demostrado que este efecto también depende en gran medida de la naturaleza del espesante empleado.

Igualmente, los resultados obtenidos en ensayos oscilatorios de baja amplitud dependen altamente del tipo de espesante, así como de la influencia de la polaridad en cada tipo de espesante. Así pues, se han observado diferentes efectos que habían sido detectados en los ensayos transitorios anteriormente realizados.

En este trabajo se han postulado y validado diferentes modelos que conciernen los mecanismos físicos por los que polaridades de los aceites base afectan el comportamiento tribológico y reológico de las grasas, así como la interdependencia de estos dos.

Este trabajo concluye con un examen riguroso de los efectos que ayudan a esclarecer los mecanismos de trabajo en contactos lubricados con grasas y que originan resultados de desgaste en todas las grasas y en las combinaciones de materiales utilizados. Uno de estos efectos se refiere al mecanismo por el que las diferentes polaridades de los aceites base influencian directamente el desgaste. El otro efecto resalta el mecanismo que actúa sobre la estructura interna de las grasas, el cual ha sido esclarecido mediante un análisis reológico. Finalmente, se ha demostrado que estos dos efectos están afectados mutuamente.

Abstract

There is a growing demand of environmentally friendly lubricating greases that presently can only be satisfied partially. The purpose of this work is to fundamentally investigate the effects that grease components exert on the resulting rheological and tribological characteristics. A limited number of different components—that is four different base oils and four structurally different thickener systems—were selected, which qualify these greases as bio-lubricants. All sixteen greases were formulated at identical NLGI grades to eliminate the consistency's influence on the results.

Polarity effects were expected to come to light during the tests. For this reason, the polarity of the matrix of different base oil candidates used for the grease formulation was determined.

These greases then underwent a series of tribological experiments on a nano tribometer test rig with the application of four different normal forces and two different material combinations resulting in individual sets of Hertzian stresses.

Additionally, atomic force microscopy observations of all model greases were carried out to give evidence to their micro structural setup.

The model greases were also rheologically investigated under different deformation situations. On the one hand, the flow response in transient shear was monitored in rotational tests, on the other hand, the greases were rheologically characterized by both, the frequency and amplitude dependent responses in small-amplitude oscillatory shear. All thus found results were also evaluated from an energetic point of view.

In addition, a series of rather unconventional test were performed with the model greases. These were tensile tests with a plate-plate rheometer where a defined gap was filled with grease and then subjected to a separation of the plates while logging the normal force and bouncing ball tests where a steel ball hit a grease-covered steel plate after having fallen a preset distance. In latter tests the resulting bouncing hight, the length of the drawn grease string as well as the impact mark were investigated.

The main findings of this work are based on influences that the different base oil polarities exerted on the tribological as well as the rheological grease characteristics.

For one thing, they were revealed in the type, severity and extent of wear marks produced in the contacting surfaces of tribological tests. Here it was shown that these base oil polarity influences highly depend on the applied thickener system and the investigated material combination. It has even shown that polarity influences can be also significantly dampened depending on the material combination.

For another thing, polarity influences were detected in the outcome of the rheological grease investigation. Here it was shown that the extend with which they come to act highly depends on the employed thickener system. It has also shown that the base oil polarity influenced the percentage thickener share depending on the type of thickener system. In the transient shear flow tests an attempt has been made to eliminate the rheological results from this thickener percentage. This is done by the evaluation of the activation energy, necessary for the structural thickener degradation revealing a large difference between greases based on highly polar oils and those based on low-polarity oils. It is shown that this effect also highly depends on the applied thickener type.

The results of small amplitude oscillatory shear tests also highly depended on the applied thickener type with still clearly evident polarity influences in each type of thickening system. Here, different effects come to act that had partially been shown in the prior performed transient shear tests.

Different models were postulated and proved in this work that pertain to the physical working mechanisms of base oil polarities and their effects on the tribological and the rheological grease behavior as well as the interdependence of these two.

This work is concluded by a disquisition of effects that help to elucidate the working mechanisms in the lubricated grease contact and which evoke individual wear results in all grease systems and material combinations. One of these effects pertains to the mechanisms that come to act with different base oil polarities and the direct influence they exert on wear results. The other effect highlights the mechanisms that come to act in the internal grease structure which has been elucidated by rheometric assessment. It is also shown that these two effects influence each other.

Contents

1. **Introduction**   1
   - 1.1. Objectives   2
   - 1.2. Structure of this work   2

2. **State of scientific research**   5
   - 2.1. Lubricating greases   5
     - 2.1.1. General remarks and definition of greases   5
     - 2.1.2. Function and application of lubricating greases   5
     - 2.1.3. Composition of lubricating greases   6
       - 2.1.3.1. Thickeners   6
       - 2.1.3.2. Base oils   10
       - 2.1.3.3. Additives in lubricating greases   12
       - 2.1.3.4. Biogreases   15
   - 2.2. Microphysical mode of action of lubricating greases   16
     - 2.2.1. Primary valence bonds   16
       - 2.2.1.1. Covalent bonds   16
       - 2.2.1.2. Ionic bonds   17
       - 2.2.1.3. Metallic bonds   17
     - 2.2.2. Secondary valence bonds—van der Waals bonds   17
       - 2.2.2.1. Keesom forces   17
       - 2.2.2.2. Debye forces   18
       - 2.2.2.3. London forces   18
     - 2.2.3. Molecular bonds in association with lubricating greases   18
       - 2.2.3.1. Determination of secondary valence bond forces in greases   18
   - 2.3. Tribology   19
     - 2.3.1. Definition of Tribology   19
     - 2.3.2. Historical review of Tribology   20
     - 2.3.3. The tribological system   20
     - 2.3.4. Surfaces of solid bodies   21
       - 2.3.4.1. Contact model of solid bodies   22
       - 2.3.4.2. Contact model of solid bodies with intermediate grease film   23
     - 2.3.5. Friction   24
       - 2.3.5.1. Friction types   25
       - 2.3.5.2. Friction states   26
       - 2.3.5.3. Friction mechanisms   29
       - 2.3.5.4. Friction energy   30

Contents

         2.3.6. Wear 33
                 2.3.6.1. Wear types 33
                 2.3.6.2. Wear states 33
                 2.3.6.3. Wear mechanisms 33
                 2.3.6.4. The energetic approach to solid state wear 35
                 2.3.6.5. The energetic approach to liquid state wear 38
                 2.3.6.6. The energetic approach to solid and liquid state wear
                          by means of entropy balance 39
  2.4. Tribometry and evaluation of tribological data 40
  2.5. Rheological tests and energetic interpretation 41
         2.5.1. Rotational transient tests 41
         2.5.2. Oscillatory shear tests – Amplitude sweeps 43
         2.5.3. Oscillatory shear tests – Frequency sweeps 47
         2.5.4. Tensile tests 48
  2.6. Other tests 49
         2.6.1. Bouncing ball tests 49

3. Assumptions and models   53
  3.1. Assumptions and deriving models for the microphysical behavior of
       greases 53
         3.1.1. Model A – physical working mechanisms of oil polarity affect-
                ing the tribological behavior 54
         3.1.2. Model B – physical working mechanisms of oil polarity affect-
                ing the rheological behavior 55
         3.1.3. Model C – interdependance between rheological and tribolog-
                ical characteristics of greases influenced by base oil polarity . 55

4. Materials, methods and conditions for the experimental analysis of greases 57
  4.1. Selection of relevant materials 57
         4.1.1. The greases 57
                 4.1.1.1. The base oils 57
                 4.1.1.2. The thickeners 58
  4.2. Selection of experimental procedures 59
         4.2.1. Polarity measurements and equipment 59
         4.2.2. AFM imaging and equipment 61
         4.2.3. Tribological tests and equipment 61
                 4.2.3.1. Tribological test parameters and materials 63
         4.2.4. Rheological tests and equipment 64
                 4.2.4.1. Rotational transient tests 64
                 4.2.4.2. Amplitude sweep tests 64
                 4.2.4.3. Frequency sweep tests 64
                 4.2.4.4. Tensile tests 65
         4.2.5. Other test equipment 65
                 4.2.5.1. Bouncing ball tests 65

5. Experimental analysis   67
  5.1. Polarity analysis 67

5.2. AFM analysis of the greases 67
5.3. Tribological analysis 73
  5.3.1. Metal soap-, HDS- and BT-greases in sapphire-steel contact . 74
    5.3.1.1. Friction results 74
    5.3.1.2. Wear results 75
    5.3.1.3. Intermediate conclusions 83
  5.3.2. Metal soap-, HDS- and BT-greases in steel-steel contact . . . 84
    5.3.2.1. Friction results 84
    5.3.2.2. Wear results 85
    5.3.2.3. Intermediate conclusions 90
  5.3.3. Comparison of tribological results with different material combinations 90
    5.3.3.1. Base oil-, thickener type- and material combination-dependent frictional and wear responses 91
    5.3.3.2. Creation of "speckle patterns" with the use of BT-greases 95
    5.3.3.3. Highly abrasive behavior of clay greases 96
    5.3.3.4. Intermediate conclusions 97
5.4. Rheological analysis 97
  5.4.1. Rotational transient tests 98
    5.4.1.1. Results and discussion of rotational transient tests . 98
    5.4.1.2. Intermediate conclusions on rotational transient tests 106
  5.4.2. Amplitude sweep tests 107
    5.4.2.1. Results and discussion of amplitude sweep tests . . 107
    5.4.2.2. Intermediate conclusions on amplitude sweep tests . 115
  5.4.3. Frequency sweep tests 117
    5.4.3.1. Results and discussion of frequency sweep tests . . . 117
    5.4.3.2. Intermediate conclusions on frequency sweep tests . 119
  5.4.4. Tensile tests 120
    5.4.4.1. Results and discussion of tensile tests 120
    5.4.4.2. Intermediate conclusions of tensile tests 122
5.5. Other tests 123
  5.5.1. Bouncing ball tests 123

6. Main conclusions 129

A. Additional information pertaining section 5.3.1 153
A.1. Interferometric investigation of selected wear marks 153

B. Additional information pertaining section 5.4.3 159

C. Scientific publications in this work 161
C.1. Journal articles derived from this work 161
C.2. Congress communications derived from this work 161

1. Introduction

Although undiscovered fossil hydrocarbon occurrences are predicted and technological progress will for some part increase the degree of feasibility of these resources we will inevitably face the limits of availability of petrochemical resources in future [1, 2]. This will require a large change of ideas in the production of auxiliary and process materials in a large number of engineering fields. Consequently, a demand is made to focus on the use of renewable materials. Moreover the steadily growing sense of ecological responsibility in society as well as legislation in concern of dealing with substances, which may have lasting harmful effects on the environment puts a challenge to producers and users of lubricants to fall back upon biologically compatible alternatives. In the background of this steadily growing awareness of ecological regards a special focus should be drawn to the twofold ecological impact of lubricants, which not only may play a superior role in the reduction of energy losses caused by friction in tribosystems of many machine elements, but also they minimize wear. In recent past the aforementioned growing awareness of ecological issues led to a growing demand of environmentally friendly lubricants. According to Wilson [3], a huge amount of lubricant leaks into the ground and sea in industrial operations such as surface mining, agricultural and constructional applications thus polluting the environment. In the last decades biological greases have been introduced into a variety of industrial and consumer applications in order to protect the environment from the negative effects of lubricants based on petrochemically refined mineral oil. Therefore, the main ecological impact of bio-lubricants, dealt with in the scope of this work, is based on their biological degradability with its resulting elimination of direct and indirect environmental pollution caused by a closed loop carbon dioxide derivation cycle. Due to these reasons, it is reported that oleochemicals have a sustainable impact on global atmospheric carbon dioxide balance [4]. It was found that, due to their lower molecular weight, vegetable-derived lubricants are significantly more degradable by maritime microbial communities than their mineral-derived lubricant counterparts [5]. In combination of these demands one can easily detect a trend towards regrowing biodegradable process materials. With respect to triblogical characteristics, vegetable oils in lubricants, in many cases, do not show any disadvantages to their synthetically or petrochemically derived counterparts [6,7]. Moreover, many of the main negative features encountered with the use of biogenic ester base oils such as oxidative, aging and thermal stability may be overcome to some degree by enzymatic or catalytic modification (transesterification, selective hydrogenation and oligomerisation) of the vegetable oil esters [6].

1. Introduction

1.1. Objectives

In view of the fact that legislation in regards with agricultural and nature exposed application of operating materials has been changing towards a more environmentally protective position in recent past, and with this process ongoing even more strictly in the future, it has been made mandatory to investigate alternative lubricants. Biodegradable raw materials are applied as supplements for petrochemical oils in these special lubricants. This, in turn, implicates some effects that are inherently integrated in the lubricants when applying biologically rapidly degradable raw materials, especially biodegradable oils. It is necessary to investigate these effects more closely in order to estimate their effects and to be able to produce better bio-lubricants. In this context, bio-greases represent a very interesting kind of bio-lubricants as the mentioned effects influence their tribological as well as their rheological behavior.

Therefore, the main objective of this work is to investigate grease behavior affected by base oils and thereby gain a deeper understanding of the influence of oil polarity on their tribological and rheological behavior. The following intermediate steps will help to reach the main objective of this work:

- Theoretical models of the tribological and rheological working mechanisms of bio-lubricating greases have to be created

- Raw materials need to be found that are suitable for the formulation of biologically rapidly degradable greases in order to achieve this goal. The variety of different raw materials has to be set to a high level with a wide band width of different polarities. This will help to attribute the found effects to the polarity influences investigated.

- Suitable experimental methods have to be selected that qualify for the detection of polarity influences in combination with the investigation of the tribological and rheological grease behavior.

- All performed series of experiments have to be evaluated in a way that the main working mechanisms of the rheological and tribological behavior of lubricating greases are disclosed, and consequently they will qualify for the verification of the theoretical models.

- Additionally, it is aimed to be clarified how the micro-rheology in the lubricant gap of tribological systems influences their tribological behavior.

1.2. Structure of this work

This work is structured in a way that first the state of scientific research in the field of lubricating greases will be outlined. In this, all components of biogenic lubricating greases are described and explained. What follows is a description of the microphysical mode of action of lubricating greases, which is substantially based on their bonding character and the bonds of contacted surfaces. Primary and secondary valence bonds are differentiated.

Moreover, the fundamentals of tribology are named, reviewed and summarized, which, initiating with a historical overview, also highlight the system approach in tribology. Furthermore, the frictional and wear types, states and mechanisms are delineated. In addition, energetic aspects in combination with friction and wear are expounded firstly for solid body friction and later for liquid friction. The latter is mainly dealt with in the section of fundamentals, which highlights rheological investigations. That section does justice to the energetic approach of rheological wear. Also, basic rheological investigations applied in lubricating grease investigation are cited.

After terminating the state of scientific research, several assumptions that are based on the knowledge expounded in the prior chapter will be established. These assumptions are used for the development of two models pertaining the tribological and the rheological behavior respectively.

Thereafter follows a chapter on the methods and conditions for experimental investigations of lubricating greases with the purpose of finding suitable methods and conditions to prove the prior assumptions. In this background, appropriate materials are selected for the formulation of model greases. Also appropriate rheological and tribological test methods are selected.

This is followed by a disquisition of a selection of investigations carried out in this work with an accompanying discussion of findings. This chapter represents the core of this work, which can not only be seen by its length but also by the number of scientific publications, in which the main findings and discussion results are presented.

This work is concluded by a discussion of principal inferences. This is done by a confrontation and comparison of tribological and rheological results.

2. State of scientific research

2.1. Lubricating greases

2.1.1. General remarks and definition of greases

Lubricating greases are defined in several different ways depending on the author or publication [8–24]. However, most of these definitions may, more or less, be summarized by the simplified statement, which defines greases as thickened lubricating oils. This statement, however, does not limit the focus of lubrication solely to the oil—the thickening agent also plays an important role in the mode of action of lubrication on both, the micro and macro levels as will be shown later on. In a more specific way, greases are lubricants, which mainly consist of a base oil, a thickening agent of either fibrous, platelet or other aggregate structure and other chemical or physical additives. The physically acting bond forces between the thickener and the base oils build a stable three-dimensional network, which has the inherent ability to regain the structure after mechanical stress. Thus, greases are non-Newtonian materials known to exhibit both solid and liquid state properties. In a lubricated contact with solid surfaces the viscoelastic properties of greases along with its tackiness allow it to remain in a spot that will be stressed tribologically in future moments or even repeatedly. On the contrary, their flow properties allow greases to yield deformational stress evoked by the relative motion of sliding or rolling contacting partners while still serving as an intermediate medium separating the two contacting surfaces on the micro scale. In addition, most lubricating greases are water resistant—a characteristic, which allows greases to work as a seal preventing water and contaminants from entering into the tribological contact. All these characteristics make greases qualified to be the lubricants of choice in a variety of applications, foremost led by rolling bearings [25, 26]. Large effort has been undertaken in the past and present research to rheologically investigate lubricating greases [14, 27–31].

2.1.2. Function and application of lubricating greases

The aforementioned characteristics of greases make clear how important each of them is for the fulfillment the lubricating greases' function and application. On the other hand, the realization of their functions helps to better understand the way lubricating greases should be designed. As indicated by the term lubricating grease its main function lies within lubrication, meaning the provision of a physically active lubricant film which separates the friction bodies and in the lasting maintenance of this lubricant film to keep it permanently stable for the prevention of direct contacts. The next main functions may be derived from these facts that is to say the active prevention of friction and wear as well as the resulting higher efficiency of the lubricated system. All functions so far mentioned analogously apply to all lubricants in general but specifically they also apply to lubricating greases, for this reason they are listed here. Very unique for lubricating greases, however, is the fact that these functions, due to their physical stability, are even fulfilled in places and surfaces to which liquid lubricants would not adhere. From this fact another important function of lubricating greases evolves—their sealing ability. With it, they actively prevent pollutants and other substances from the exterior, especially water, to come into contact with the friction spot. Especially because lubricating greases remain in the lubricated spot and adjacent regions they, more than other lubricants, protect the friction spot from corrosion. Based on their solid characteristics and a consequent force necessary for removal, lubricating greases better than other lubricants work as vibration and sound dampers.

2.1.3. Composition of lubricating greases

Lubricating greases are produced as complex multiphase systems in which all component properties are needed for fulfillment of the desired function. In order to comply with all desired functions base oils as well as a thickening agent are needed, the latter of which may take a share of up to 50 % and more depending on the required consistency and all entailed characteristics of the grease. Just as it happens with all oil based lubricants the desired tribological characteristics are only enhanced by lubricant additives. These may take a share of up to 10 %.

2.1.3.1. Thickeners

Usually, lubricating greases are classified either by industrial application or by ingredients. From a chemical point of view the classification according to thickening agents is most common. Generally the thickening agents are divided into the three groups of plain-soap-thickeners, complex-soap-thickeners and non-soap-thickeners according to figure 2.1.

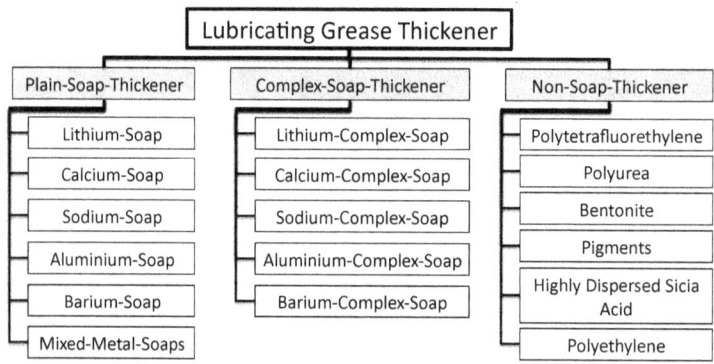

Figure 2.1.: Common thickeners in lubricating greases

Soap-thickeners Formally soaps result from the hydrolysis of an ester and the aqueous solution of a hydroxide. Products of this saponification process are the alcohol and the salt of the acid which the ester contained of. In lubricating greases these esters are usually derived from the carboxylic acids of animal or vegetable fatty acids. From a chemical point of view soaps are the salts of these fatty acids. In lubricating greases mostly metal salts of fatty acids are employed. These may be salts of either lithium, calcium, sodium, aluminum, barium, lead or other metals. All these metals form soaps of different structures, which lead to specific characteristics in the application of the lubricating grease. The presence of a metallic soap as a thickener does not only define the consistency of the grease but it also influences its frictional properties [24, 32–35].

In many cases, soaps evolve as fibrous crystalline structures of different fiber lengths. In the end of the first third of the last century theories of fibrous soap structures were confirmed with the invention of scanning electron microscopes (SEM), and their application in grease investigations [36]. With this new method of imaging investigation theories arose concerning the orientation of the molecules in the soap fibers [37]. Figure 2.2 depicts SEM images illustrating the different fiber lengths of various lubricating greases.

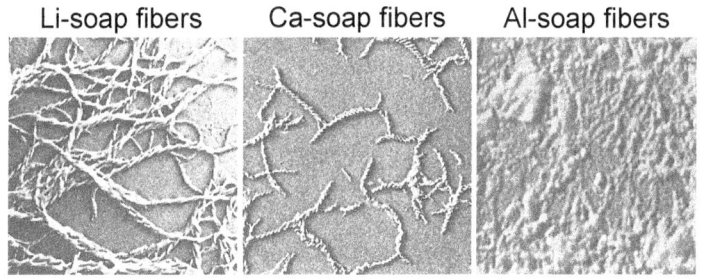

Figure 2.2.: SEM images of Fibers of various lubricating greases as taken from [38] in [36]—images taken with similar scaling, not clearly specified therein

A size comparison between single soap fibers and various viruses and bacteria made by Farrington [38] clearly highlights typical structures of single metal soaps as well as their lengths (see figure 2.3).

Calcium soap Calcium greases nowadays are usually formulated from calcium hydroxide and stearic acid. The utilized fatty acids for the formulation may be of plant and animal origin. Calcium greases are characterized by a short fiber structure, and they tend to dropping points at approximately 100 °C while their long term application temperature should not exceed 80 °C. With the application of Ca-12-hydroxy stearate their dropping point is raised to 130 °C approximately with a recommended operating temperature of no more than 120 °C. Calcium greases in general are very water-resistant and show good adhesion characteristics. They are often used as a simple machinery lubricating greases and because of their relatively

2. State of scientific research

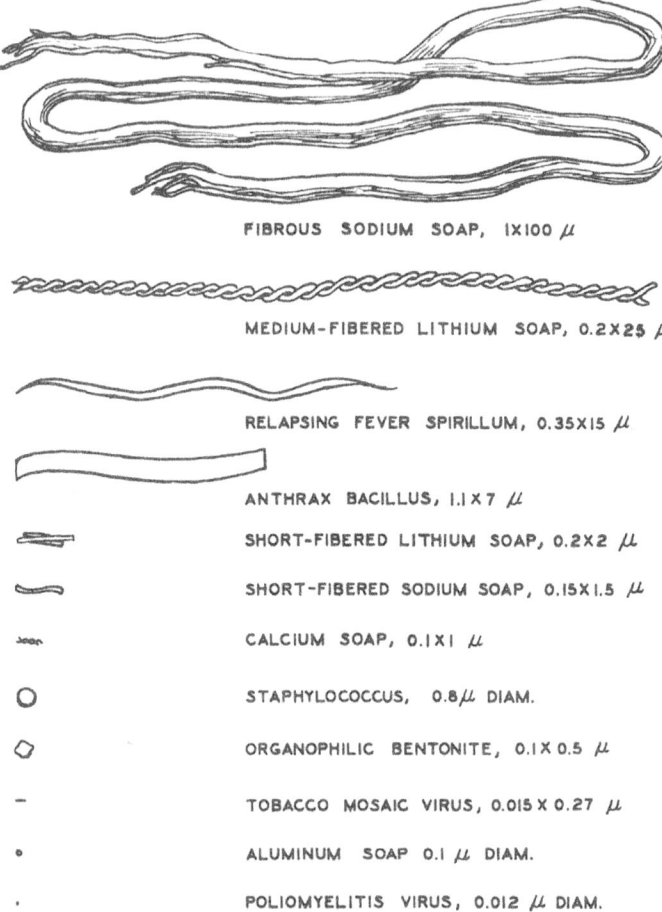

Figure 2.3.: size comparison between single soap fibers and various viruses and bacteria as taken from [38] in [36]

2.1. Lubricating greases

low raw material costs they are often applied in steel plants, where they are used in loss lubrication. Moreover, they are also used in automotive applications.

Lithium soap Lithium soap is the most frequently used grease thickener [39] not only in the roller bearing industry. Mostly Li-greases are produced as Li-12-hydroxy stearate greases. With a dropping point of about 180 °C their recommended operating temperature, too, is much higher than that of Ca greases. Moreover, Li greases are water resistant to some extent. They also perform very well in working stability, a measure of the resistance to softening, and show good adhesive characteristics with their relatively long soap fibers. They are applied as multipurpose greases in almost all industrial application fields.

Sodium soap Sodium greases are mostly produced from sodium hydroxide and 12-hydroxy stearic acid. Although their dropping points are around 170 °C, they should not be exposed to temperature conditions higher than 120 °C for a prolonged time. Their structure is very rough and consists of long fibers. Although they are not water resistant they still offer quite good corrosion protection if the amount of water leaked into the system is not too high. This is based on the fact that introduced small amounts of water are very quickly chemically bound before reaching the tribocontact. Sodium greases are very stable against shear, exhibit good adhesive characteristics and are mostly used as flowing greases in the lubrication of gears.

Complex-soap-thickeners In opposition to simple soap greases, in which only one fatty acid or a combination of similar fatty acids are used in the saponification process, complex greases are composed of specific mixtures of long chain fatty acids and short chain carboxylic acids or inorganic acids. These complexes impart very good characteristics to the thickening agent, which mostly exhibits much higher temperature resistance than the respective regular soaps. The more complex formulation processes and the targeted better characteristics mostly result in higher prices of complex-soap-thickened greases than in regular soap greases.

Non-soap-thickeners In non-soap-thickeners other organic or inorganic material is used as thickening agents, which contribute a colloidal structure to the system by adsorption and some type of chemical bonding of the thickener particles according to Boner [36]. These can either be minerals derived from natural clays or polymers, which both exhibit inherent swelling characteristics such as bentonites or highly dispersed silica acid and polyethylene or polyureas, respectively.

Bentonite (BT) Bentonite is a clay-based thickener extracted from clays of volcanic ashes with montmorillonite, a triple-layer particle structure phyllosilicate, as main constituent. The typical layer structure is well displayed in figure 2.4 taken from [40]. Montmorillonite itself chemically consists of 60 % SiO_2 and 20 % Al_2O_3 and contains several other metal ions stored in the layer lattice [41–43]. Substitution of these ions by quarternary ammonium salts results in organophillic modification, which enables thickening processes of petrochemical as well as oleochemical oils [41, 44–47]. As a

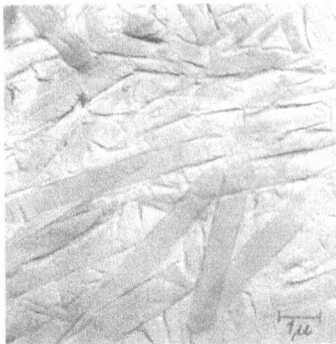

Figure 2.4.: Montmorillonite layers in scanning electron microscopy taken from [40]

result of this modification the surface of the montmorillonite particles is activated with hydrocarbon alkyls of different chain lengths. The length and polarity of these chains influences the affinity for oils according to their polarity. In the thickening process intense shearing is needed in the process of blending the base oil with the organically modified bentonite in order to break up the plate layer aggregates and completely wet all surfaces. Addition of other chemical activators (carbonates, glycols, ketones, alcohols and especially water) is needed to overcome adhesion forces between the particles and enable the thickening process. During the latter, the particles form a three-dimensional stable network, which physically traps the base oil [7, 41].

Highly-dispersed silica acid (HDS) Due to its main constituent, SiO_2, HDS-thickener, like BT is regarded as a clay thickener. In the production process, quartz sands are firstly enriched with chlorine resulting in $SiCl_4$, which is then pyrolytically cracked in oxyhydrogen-reactive flame hydrolysis resulting in SiO_2 primary particles consisting of tetrahedral SiO_4 connected via siloxane bridges and outstanding silanol groups. These primary particles link together in the flame reaction and form flaky chain aggregates with a very high specific surface (up to $400 \, m^2/g$). The silanol groups of the primary particles enable hydrogen bonds to organosilicon compounds in the hydrophobization process and to the hydrocarbon chains of surrounding base oils in the thickening process. Through the latter, a three-dimensional network structure is formed that provides mechanical stability [41].

2.1.3.2. Base oils

Petrochemical base oils Nowadays, most lubricant base oils are refined of crude oil by various processes. Crude oil consists of lots of different hydrocarbons of varying chain lengths and structures (aromatic, naphthenic or paraffinic). In the refinery process the crude oil is distilled into different fractions, which results in a sorting of hydrocarbons by chain lengths. Dewaxing, hydrofinishing, catalytic hydrocracking, cracking, synthesis and hydrogenation are subsequent processes that further refine these intermediate products into final products such as raffinates, synthetic hydrocarbons, hydrocrack oils and esters, each with specific characteristics depending on the production process. All quoted products may be used as base stocks for lubricants. Characteristic descriptions of base oils such as synthetic, fully-synthetic, parasynthetic, semisynthetic, synthetic-based and many others are mere terms of marketing that lack a standardized labeling [48]. Although the classification of lubricant base stocks is not standardized in a norm, the classification according to the American Petroleum Institute (API) into groups I through V has established itself as a quasi standard in the lubrication industry. In API groups I through III the only criteria for differentiation are the degree of saturation, the sulfur content and the viscosity index of the base oils. While API group IV lists synthetic hydrocarbons such as polyalphaolefins, API group V lists all base oils that are not listed in groups I through IV—mainly synthetic oils such as esters, polyglycols, silicon oils and others.

From an economic point of view it should be mentioned that base oil prices rise along with their API group assignment. The more chemically complex the manufacturing processes are, the more expensive are the end products. These high prices may very well be justified by more stable tribological load carrying performance, chemical and thermal stability and other substantial advantages of these base oils compared to their cheaper counterparts in lower API groups. With the current trend of rising physical requirements to lubricants a general trend towards higher degrees of synthesizing may be detected.

Biogenic base oils Throughout most parts of mankind history oils derived from naturally regrowing resources have been used as lubricants. With the beginning of the age of industrialization mineral oils were discovered and petrochemicals replaced these natural products [7]. Especially in the sense of a growing ecological awareness and with respect to limited petrochemical resources as described in the introduction of this work it is necessary to start a rethinking process back to biological base stocks. These biological fats originating from vegetable, marine or animal resources such as rapeseed, palms, sunflower seeds and many others qualify very well as base stocks for the production of bio-lubricants. Mostly they are constituted of triglycerides, which are esters of long chain fatty acids and glycerol. As all these are natural products they do vary in composition depending on the biosyntheses or the metabolism of the originating organism. It is reported that some of these biological oils contain up to twelve different fatty acids [7]. Biological esters are chemically generated by the reaction of an alcohol (R'-OH) and an acid (R-COOH) under the cleavage of water [7,49] as depicted in figure 2.5. Esters always comprise of two fundamental

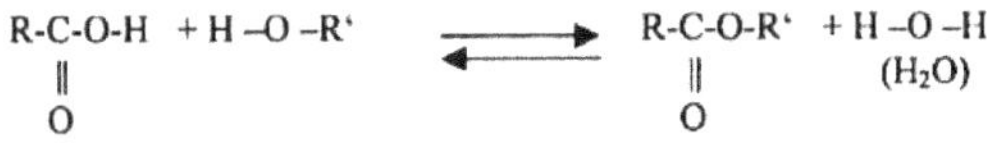

$$\text{R-C-O-H} + \text{H-O-R'} \rightleftharpoons \text{R-C-O-R'} + \text{H-O-H} \quad (\text{H}_2\text{O})$$
$$\|\qquad\qquad\qquad\qquad\quad\|$$
$$\text{O}\qquad\qquad\qquad\qquad\quad\text{O}$$

Figure 2.5.: Ester reaction in oleochemicals as found in [49]

elements, the C=O double bond (carbonyl group) and the single bond oxygen in the mutual carbon atom. The carbonyl group of esters with its high difference in electronegativity is what causes their high polarity. The most frequent fatty acids in biological esters are listed in table 2.1.

Double bonds in the fatty acid chains of esters are points of instability which offer targets to chemical reactions such as oxidative processes. Therefore the number of double bonds, which is a degree of unsaturation, marks the chemical instability of biological esters.

As the types of fatty acids and the alcohol with its accompanying branching in esters substantially influence their tribological characteristics as well as their physicochemical properties such as viscosity, viscosity index, pour point and thermal stability it is important to keep these types at a constant level throughout the complete ester. Therefore natural esters are often refined and synthesized to attain taylor made characteristics. Especially with regards to this trend of a higher degree of synthesis in the production of lubricants the application of regrowing base stocks for their production processes becomes more and more reasonable. They qualify for synthesizing processes, especially esterification, just as well as petrochemical refinery intermediate products do.

Chemical stability, especially oxidative stability, needs to be taken care of very well with these oleochemical products but also with petrochemical products. But the procedures of enzymatic and catalytic modification, mentioned in the introduction of this work, result in base oil esters on a regrowing basis that are just as stable as their petrochemical counterparts [6].

2.1.3.3. Additives in lubricating greases

Generally speaking, lubricating grease additives are chemical substances that provide the grease with desired specific characteristics, which are not or only insufficiently inherently present in the base oil mix. The definition given by Webber [50] probably most properly describes the function of additives:

> Additives are substances which impart or enhance the desirable properties of a lubricant and eliminate or minimize the deleterious ones.

Additives are often differentiated by their targeted influence. There are additives, e.g. which affect the mechanical stability of the thickener matrix, the oxidative stability of the grease, their frictional and wear behavior, their tackiness or their protection of tribological surfaces against corrosion. Other authors [51] differentiate additives by their mode of action into surface-active [52, 53] and fluid-stabilizing additives. Additive packages applied in lubricating greases generally do not differ substantially from those applied in solely fluid lubricants. The most crucial feature to be considered in the application of additives, in combination with all lubricants and especially greases, is their possible mutual interference. For this reason the additive package needs to be well balanced in relation to compatibility with all grease components.

Oxidation inhibitors Hydrocarbons are most susceptible to autoxidation mechanisms [54–56] which may greatly be accelerated by the presence of catalytically

Table 2.1.: Most frequent fatty acids in biological esters with comprising fatty acids according to [7]

Natural source	Fatty acid type	Carbon chain length and number of double bonds
Soya bean oil	Linoleic	C18:2
	Oleic	C18:1
Groundnut oil	Linoleic	C18:2
	Oleic	C18:1
Palm oil	Palmitic	C16:0
	Oleic	C18:1
Rapeseed oil	Oleic	C18:1
	Erucic	C22:1
	Linoleic	C18:2
Sunflower oil	Oleic	C18:1
	Stearic	C22:1
	Linoleic	C18:2
	Palmitic	C16:0
Groundnut oil	Oleic	C18:1
	Linoleic	C18:2
Beef tallow	Oleic	C18:1
	Palmitic	C16:0
	Stearic	C18:0
Lard	Oleic	C18:1
	Palmitic	C16:0
Coconut oil	Lauric	C12:1
Palm kernel oil	Lauric	C12:1
Olive oil	Lauric	C18:1
Fish oil	Long chain	C20:2 to 6
	Fatty acids	C22:2 to 6
Corn oil	Linoleic	C18:2
	Oleic	C18:1
Cotton seed	Oleic	C18:1
	Palmitic	C16:0
	Linoleic	C18:2
Castor oil	Ricinoleic	C18:1-OH
Linseed oil	Linolenic	C18:3
Crambe oil	Erucic	C22:1

working metal ions, high temperatures and mechanical shearing. These mechanisms include a free radical chain reaction [57] and can be inhibited if the radicals are scavenged by antioxidants such as listed in [57]:

- Hindered phenols
- Aromatic amines
- Metal dialkyldithiophosphates
- Metal dialkyldithiocarbamates
- Ashless dialkyldithiocarbamates
- Sulfurized phenols
- Phenothiazine
- Disulfides
- Trialkyl and triaryl phosphates and phosphites

Extreme-pressure and anti-wear additives The functionality of most extreme-pressure and anti-wear additives is based upon surface layer formation caused by surface activity. Molecules dock with their polar ends to tribological surfaces where they form stable protective layers. This protection is especially crucial in mixed state friction regimes. Examples of such physisorptive additives are taken from [57]:

- Sulfurized olefins, fats, and esters
- Chlorinated paraffins
- Metal dialkyldithiophosphates
- Phosphate and thiophosphate esters
- Ammonium salts of phosphate esters
- Borate esters
- Metal dithiocarbamates
- Metal naphthenates
- Metalsoaps
- Sulfides and disulfides
- High-molecular-weight complex esters

Rust and corrosion inhibitors Since most technical surfaces considered for lubrication consist of ferrous metals it seams obvious that these are subject to natural oxidation processes. In this focus, rust is developed as a result of an electrochemical process, in which iron and atmospheric oxygen interact. The forming of this process may be accelerated by the catalytic action of water [58]. Rust and corrosion inhibitors act in different ways to prevent oxidation from occurring on tribological surfaces. While corrosion inhibitors neutralize acids, which may develop from the degradation of hydrocarbons rust inhibitors are targeted to counteract the harmful effects of water by chemically binding water on the one hand, and by forming protective layers impermeable to water on the tribosurface on the other hand. This is mainly realized by polar additives docking to surfaces by means of physisorption. This background reveals the partially antagonizing effects of different additive types. Extreme-pressure and anti-wear additives also work by physisorbitve adhesion to surfaces, as has been mentioned earlier. This may result in a conflict if the dosage of respective additives is not well balanced. Typical rust and corrosion inhibiting additives are listed in [57]:

- Carboxylic acids
- Salts of fatty acids and amines
- Succinates
- Fatty amines and amides
- Metal sulfonates
- Metal naphthenates
- Metal phenolates
- Nitrogen-containing heterocyclic compounds
- Amine phosphates
- Salts of phosphate esters

2.1.3.4. Biogreases

When formulating greases, which fully meet both above-mentioned requirements of biogeneity and biodegradability, the manufacturer is challenged to take all of this into consideration for the selection of all ingredients. If, for reasons of feasibility and lack of experience, not all components may be replaced by such of biogenic origin the utilization of biogenic base oil may at least be a step into the right direction. Hence, the replacement of mineral derived base oils by such of naturally regrowing and therefore rapidly biodegradable origin, in order to produce bio-lubricating greases, has been attempted in many cases. These substituting oils or oleochemicals are mostly refined from vegetable oils. Since oleochemicals intrinsically exhibit positive tribological properties these substituted greases must not inevitably be inferior to the original greases based on mineral oils. This approach seems to be obvious when considering the composition of greases with the base oil taking a share of up to 95 % depending on the targeted grease consistency and application field, as described earlier. With regards to conformity to the demands of eco labels like e.g. the German Blue Angel, however, the mere substitution of the base oil with a rapidly biodegradable one is still insufficient. Also for the thickening agent, which is regarded a basic substance, biodegradability must be verified by at least 70 % using one of the test methods according to OECD 301 B, F, D or C, ISO 14593 or ISO 10708 [59]. Although biodegradable thickeners are scarcely found in current presence on the world market, previous investigations conducted in the field of biogenic thickening agents have been reported. Thus, the use of different biopolymers derived from natural constituents such as polysaccharides as thickening agents in the formulation of biologically and environmentally acceptable lubricating greases has been proposed [59]. In this sense, some formulations containing thickening agents derived from cellulose and chitin/chitosan and castor oil have been rheologically [60–62] and tribologically [63] characterized. The basic criteria for award of the Blue Angel environmental label also accept mineral thickeners in bio-greases.

2.2. Microphysical mode of action of lubricating greases

If in tribological investigations one tries to comprehend how solid bodies and matter in general interact with each other, one needs to realize first how this matter is constituted and what makes the inner cohesion forces work. One needs to also keep in mind that the inner bond forces of matter substantially influence the material's propensity to interact beyond its boundaries. In this context the two solid bodies in tribological contact play the same role, which is subordinate to the role of the liquid intermediate material. This superior role of the lubricant enables its function—the active, mechanical separation of the solid bodies even under the effect of tribological load. The cohesion of material and the chemical mode of action of lubricants, and greases in particular, are based on different types of bonding, which will be explained below. Whether valence electrons of the outer atomic shells are exchanged or not, differentiates if these bonding types are primary or secondary valence bonds.

2.2.1. Primary valence bonds

Atoms exchange valence electrons in the bond structure of primary valence bonds. The bond strength of primary valence bonds results from the mode of action of atomic electron exchange. According to the nature of this electron exchange a distinction is made between covalent, ionic and metallic bonds.

2.2.1.1. Covalent bonds

According to the octet rule, atoms seek to reach noble gas configuration, in which the outer atomic shells contain eight electrons. For this reason atoms share electrons within a bonding structure in a way that possibly all participating atoms reach this configuration. If the bond structure of molecules or crystals is based on such kind of electron exchange it is called a covalent bond. Covalent bonds are the strongest chemical bonds. In the exchange of valence electrons different electron affinities of atoms or molecules may result in charge displacements. This results in certain charge imbalances, which may evoke a polar character of the structure. The electronegativity is a frequently used measure of this dipole character [51, 64].

2.2.1.2. Ionic bonds

The polar character of a covalent bond increases along with its electronegativity. A bond is called ionic bond if its electronegativity is large enough to completely push single electrons from the outer shell of one atom to another. The transition from covalent to ionic bonds is gradual. In single cases, the ionic character of a bond may be strong enough to remove all electrons from an outer atomic shell.

2.2.1.3. Metallic bonds

Metallic bonds are characterized by very low electronegativity, which results in electrons that can freely move in the crystal lattice. The outstanding electrical conductivity of all metals is based on this so-called electron gas. Metallic bonds are the weakest of all primary valence bonds.

2.2.2. Secondary valence bonds—van der Waals bonds

Although secondary valence bonds are much weaker than primary valence bonds they play the most important role in the mode of action of lubricants. The substructures of lubricating greases on the molecular level, namely metal soaps of thickeners, which are referred to as of polymeric structure [36] and the molecules of hydrocarbon chains in base oils, are all constituted of primary valence bonds. Yet all such lubricants adhere to tribosurfaces and to each other exclusively by means of secondary valence bonds. Consequently all tribological and rheological interactions of lubricants are based on secondary valence bonds.

Secondary valence bonds act because of attracting forces between molecules with dipoles of different strengths. These so-called van-der-Waals bonds rest upon electrostatic interaction between molecules with permanent or temporary dipoles [65] in [51]. Depending on the dipole character of the molecules involved it is differentiated between Keesom, Debye, and London forces. [66]

2.2.2.1. Keesom forces

Keesom forces are the strongest secondary valence bonds. They come to act when two molecules with permanent dipoles interact with each other. These molecules align according to their polarity and the resulting electrostatic attraction. So, if for any reason alignment of these molecules is inhibited they could only distract each other. Intermolecular movement or oscillation due to temperature effects does have such an inhibiting effect on the alignment, therefore the formation of Keesom forces is highly temperature dependent. Hydrogen bridge linkages constitute a special case of Keesom forces, in which a single hydrogen atom bridges two molecules with permanent dipoles [51, 58, 64, 67].

2.2.2.2. Debye forces

Debye forces result in the attraction of one molecule with a permanent dipole and one molecule without permanent dipole. The electrostatic forces of the dipole molecules induce a dipole momentum in the other ones, which results in attracting forces [51, 58, 64].

2.2.2.3. London forces

London forces rest upon electron movement within covalent bonds of two adjacent molecules without permanent dipoles. The electron movement of one molecule induces a temporary dipole in the adjacent one, which results in mutual attraction. London forces are the most frequent and also the weakest secondary valence bonds. In the active functionality of friction and wear inhibiting components of lubricants they only play a minor role [51, 58, 67].

2.2.3. Molecular bonds in association with lubricating greases

If considering the mass distribution and the mixing ratio of thickener and oil in lubricating greases there must be a physical reason why the relatively small amount of thickener is able to thicken such a large proportion of oil. Bondi et al. [37] attribute this fact to capillary action in smallest interstices between thickener structures. Boner [36] also refers to the conclusions of Lawrence and McBain et al. [68, 69], namely the sorption of oil molecules to protruding, attracting groups in the thickener structure and the penetration of free oil molecules into the interwoven thickener network resulting in an energetically more favorable level of the network. The latter is concluded to be proven by the swelling of the thickener network. In the case of some thickener structures a chemical activation of the thickener particle surface is necessary to promote the thickening mechanism. This, however, strongly indicates to an interaction of oil polarities and the thickener structure. In other words the base oil adheres to the thickener structure owed to secondary valence bonds. For further tribological and rheological discussions included in this work it will be important to determine the strength of these adhesion forces.

2.2.3.1. Determination of secondary valence bond forces in greases

The magnitude of secondary valence bond forces, mainly of the base oils, needs to be determined to draw inferences about the bonding character between the oil and the thickener structure within a grease. This defines its cohesion forces. Also secondary bond forces of the oils have to be determined to draw conclusions about the grease's propensity to adhere to tribological surfaces. In other words the base oil's bond forces are the physical properties that define the cohesion and adhesion character of greases. These forces, however, cannot be determined by measurements but only by theoretical considerations. Massmann [51] attributes the van-der-Waals-interactions to the dipole momentums, which largely depend on the difference of electronegativities. According to this approach, dipole momentums exert the largest influence on van-der-Waals-interactions defined in [70] as

2.3. Tribology

$$w_{vdW}(r,T) = -C \cdot r^{-6} \quad (2.1)$$

where r is the distance between the atoms, T is the temperature and C is equal to the sum of all van-der-Waals-interaction energies [51]. This is caused by the influence which the dipole momentums exert on van-der-Waals-interaction energies, as can be seen in

$$C = C_{Keesom} + C_{Debye} + C_{London} = \frac{\frac{D_i^2 + D_j^2}{3kT} + (D_i^2 \alpha_{0j} + D_j^2 \alpha_{0i}) + \frac{3}{2}(\frac{\alpha_{0i}\alpha_{0j}h\nu_i\nu_j}{\nu_i+\nu_j})}{4\pi\epsilon_0^2} \quad (2.2)$$

where i and j are the indexes of the molecules, D is the dipole momentum, ϵ_0 is the electric field constant, h is the Planck constant, k is the Boltzmann constant, α_0 is the polarizability, T is the temperature and ν is the excitation frequency [51].

Furthermore it is deduced [51] that the dipole momentums, defined as the product of the charge (q) and the distance of the charge (d),

$$D = q \cdot d \quad (2.3)$$

can only be determined with very complex measurements. Therefore, Massmann [51] introduces a *'relative adsorptivity'*, $A_{molecule}$, as a way to simply estimate the propensity of polar-active molecules to adsorb to metal surfaces. It is defined as the ratio between the product of the number of bonds $\sum N_{AB}$ between two elements A and B of that molecule and the difference of electronegativities ΔX_{AB} and the number of molecule branches of that specific molecule, $M_{molecule}$

$$A_{molecule} = \frac{\sum N_{AB} \cdot \Delta X_{AB}}{M_{molecule}} \quad (2.4)$$

As well as this approach works, it needs the precise knowledge of electronegativities and the number of bonds and branches within a molecule. If, however, this knowledge is not available, other ways have to be found to estimate the magnitude of dipole momentums in a given oil molecule. The dipole momentums are a measure of the polarity of oils. Therefore the determination of oil polarities allows for conclusions to be drawn about the size of its dipole momentums and its consequent adsorptive and cohesive behavior in combination with grease formulation.

2.3. Tribology

2.3.1. Definition of Tribology

The modern generic word 'tribology' originates from the greek verb $\tau\rho\iota\beta\omega$ 'to rub' and the noun $\lambda o\gamma o\varsigma$ 'word, science, teachings or doctrine'. It was first used in the Jost report, which came forth from a 1964 government commissioned survey on the economic losses evoked by untoward effects of friction. The committee concluded in the immense scientific relevance of tribology, which deals with the mutual interaction of solid contacting surfaces in relative motion. Tribology mainly focuses on the scientific investigation of all phenomena evoked by the three sub disciplines friction, wear and lubrication. [71]

2.3.2. Historical review of Tribology

The phenomena of friction made manifest by the resistance especially of heavy solid bodies, e.g. stone blocks, have been known to mankind for thousands of years. This may be derived from documented inventions for the reduction of frictional resistances. The invention of the wheel, not the least of which, is owed to the fact that rolling processes offer much less resistance to relative motion than sliding processes in the transportation of mentioned objects. If objects were to be moved whose weight made the use of spoke wheels impossible from a stability perspective, then massive rolling bodies such as wooden trunks were placed underneath these objects, which were manually replaced before them after a complete rollover. This mode of action is still remotely fundamental in modern Roller bearings. But also in this example a tribological system approach had to be performed, which did not only take the transportation load into consideration but also the characteristics and the consistency of the ground surface. In egyptian tomb paintings the manpowered transportation of statues weighing several tons and the application of a suitable lubricant has been made manifest even before 3000 B.C. [71]. These two examples make clear that lubrication as well as the application of mechanical load carrying elements need to be considered even in the historical disquisition of tribology. Also wear as the third principle of tribology has been known to mankind for a long time as evidenced in historical reports and shown in the following example. The wheel hubs of roman and greek chariots were presumably lubricated with vegetable oils or animal tallow in order to minimize wear and friction [72, 73]. A list of lubricative as well as mechanical means for the reduction of friction and wear could easily be expanded to several pages.

2.3.3. The tribological system

The system concept is frequently used in tribological investigations because it helps to classify discrete friction points and to delimit them from others. The use of this concept needs to be clarified first. In the system analysis according to [74], material systems are referred to as enclosed natural or artificial structures, which are composed of discrete numbers of elements. This use helps to regard tribological systems as a black box with clearly set boundaries. Systems generally have a certain structure. This structure refers to the composition of the system of individual elements. In addition, systems have an environment. This indicates to a clear distinction from other sub-systems. Furthermore, a system receives specific inputs from its environment. These input quantities can either be of material (material or substance flow) or non-material (information or energy flow) nature. The received material and energy flows are transformed or processed in the interior of a system. The resulting material and non-material flows may be conveyed out of the system. This property enables systems to send output signals to its surroundings. In the discrete application of tribology, systems always embody a discrete frictional contact

2.3. Tribology

which, according to [74] is always constituted of at least two frictional bodies and/or an intermediate medium as depicted in figure 2.6. The boundary of a tribological system is defined by its delineation to its environment, which always contains the ambient medium of the friction point.

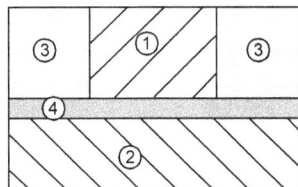

Figure 2.6.: Basic structure of a tribological system following [74]—with 1) Friction body 1, 2) Friction body 2, 3) environment and 4) intermediate medium

All influences working from the exterior of a frictional point define the input of a tribological system. These include, first and foremost, the externally applied forces, which on the one hand press the frictional bodies against each other and, on the other hand, cause the relative motion between the frictional bodies. Depending on the type of lubrication, a lubricant flow can also be considered as an input stream. The output is defined as the mass flow of ejected wear particles on the one hand and the radiated energy of the system on the other hand. This is made manifest through a heat flow but also by an acoustic signal caused by oscillations in the friction point. The friction process that takes place in a friction point can generally be interpreted as a process of energy conversion. In other words, friction equals energy conversion in the tribological system. Any irreversible changes, as system responses, are summarized by the term wear. Or in other words, all system responses measurable in form of wear can be attributed to energy conversion. This defines a clear principle of cause and effect. The individual terms of friction and wear will be explained in more detail later on.

2.3.4. Surfaces of solid bodies

Since solid bodies can only come into contact through their surfaces it seems plausible to take a closer look at the physical properties of solid surfaces. Many factors are involved at the shaping and contouring of solid matter and each affects the constitution of the surfaces, among these, the technical process with all its tools and processing steps that are used to create the surface. It stands to reason that technical surfaces created with a cast mold process, in which the thermally fluidized base material is cast into a form where it consolidates and copies its shape, will result in different surface properties than a surface created with a cutting process. In addition, the properties of surfaces created by cutting processes significantly depend on properties and the geometry of the tool, the processing speed, the applied metalworking fluid and the basic material. All these technical processes influence the surface quality meaning the roughness and the waviness as classified in DIN EN ISO 4287.

Apart from the technical production processes also solid-state physical and chemical properties of the basic material play a significant role in the formation of surfaces. If a surface is created by a fracture mechanical process the basic material will always begin to flow at its slip planes and lattice borders, the properties and sizes of which will inevitably define the resulting surface quality. Moreover it should be noticed that the liberated bonding energy in the site of fracture will lead to an imbalance toward the bonding energies in the inner basic material. These imbalances will initiate a structural reordering on the surface [75].

Another important aspect in the creation of solid material surfaces results from chemical activity. Bonding energies liberated in the cutting process are available for chemical interactions with the surrounding media. In this context chemisorptive and phsysisorptive processes as well as chemical reactions such as oxidative processes should be mentioned. The resulting surface quality may be modeled according to [76] as depicted in figure 2.7. This figure fundamentally differentiates between the

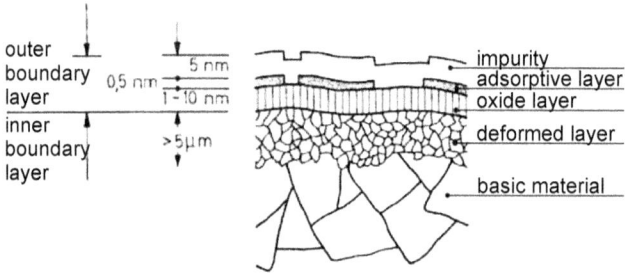

Figure 2.7.: Cross-section of a metallic surface following [76] as found in [75]

inner and the outer boundary layers, which both synergistically separate the basic material from the ambient medium. Both, the outer parts of the basic material and the deformed layer are part of the inner boundary layer. The outer boundary layer is constituted of an oxide layer, an adsorptive layer and impurities that naturally adhere to surfaces. It should be noted that this figure merely represents a model of technical surfaces with all formed layers, which does not raise claims to correct scaling. But still it highlights the present layers quite well.

2.3.4.1. Contact model of solid bodies

As explained in the previous section, tribological surfaces are never completely smooth on the micro scale. They are rather spotted with asperities, which have a substantial impact on the contact behavior of solid surfaces. Even though solid bodies seem to establish contact in the complete macro contact, in reality the contact situation is much different, as modeled in figure 2.8 [75]. The real contact surface of contacting bodies is much smaller than the nominal contact surface, as can be seen in the same figure. As each contacting asperity may be deformed to an individual

2.3. Tribology

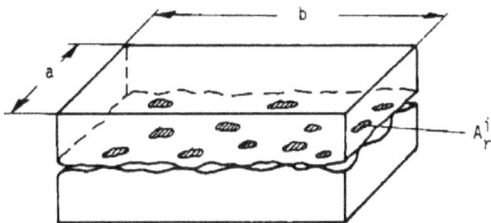

Figure 2.8.: Model of a real contact situation taken from [75]

degree it becomes clear that the real contacting area of solid bodies substantially depends on the jacking force between those bodies, their degree of hardness and other material characteristics. At this point, certain models are used, which shall help to determine the size of the real contacting surface. These models depend on the roughness and the hardness of contacting bodies. According to Fleischer et al. [77], the contacting cases in rough/smooth contact situation are to be set equal to a hard/soft surface combination. Analogously, a combination of rough/rough surfaces is to be set equal to a hard/hard contact. Moreover, Fleischer et al. [77] suggest a way to approximate roughness asperities by spherical segments, the hight and radii of which are two-dimensionally normally distributed as random variables on the surface. The contact probability is defined by F, where ξ and and ρ are random variables for the relative hight and radius of the segments, respectively.

$$F(z,u) = \int\limits_0^z \int\limits_0^u f(\xi)f(\rho)d\xi d\rho \qquad (2.5)$$

The size of the real contacting surface is estimated by reference to this normal distribution in combination with the material hardness and the consequent elastic and plastic deformation and in dependence of the applied contact model.

2.3.4.2. Contact model of solid bodies with intermediate grease film

The just-described model of solid friction is now enhanced by the idea of a separating lubricating grease film in between the solid bodies. Kuhn states that the friction process is discretized to the actual contacting areas of two opposing asperities [24]. The largest portion of energy consumption in the friction process of grease lubricated contacts is constituted of the energy density necessary for the deformation of solid surface asperities. Therefore, the mechanisms that take place inside the lubricant film are the core of investigation and modeling in combination with the contact model of lubricated friction pairs. According to Holweger [78], the separating lubricating grease film features inhomogeneities that are defined as locally differing characteristics of the lubricating grease such as density differences, flow energy density differences, individual and unique flow properties and also unevenly distributed constituents in the lubricating grease. Microscopically, greases must exhibit different attributes as they are constituted of base oils and thickeners of completely divers structures. Kuhn

2. State of scientific research

takes the Newtonian layer model as a basis of his considerations of the microscopic investigations of the frictional process in the lubricant film since this process is most exactly defined as a shear process. Consequently he attributes the just-mentioned inhomogeneities to locally limited areas of the respective grease layers as can be seen in figure 2.9. This figure shows a discrete contacting spot that is separated by a

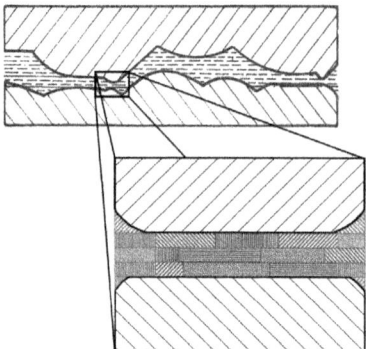

Figure 2.9.: Model of a grease lubricated contact situation following [24]

lubricating grease. With further magnification the characteristically different areas become visible. He concludes that the shear process in the micro contact depends on the distributed differences of characteristics in the grease. Thus, each regional inhomogeneity exerts its own influence on the tribological system. Consequently, Kuhn [24] introduces a third random variable, δ, in his contact model of solid bodies with intermediate grease layer that distributes these regions of inhomogeneity representing the results of IR-microscopic investigation performed by Holweger [78]. Consequently, the contact probability follows as

$$F(z,u,\rho) = \int_0^z \int_0^{z-1} f(\xi_1)f(\xi_2)d\xi_1 d\xi_2 \int_{u_k}^{u_g} f(\rho_1)d\rho_1 \int_{u_k}^{u_g} f(\rho_2)d\rho_2 \int_{\rho_2}^{\rho_1} f(\delta_1)d\delta_1 \int_{\rho_2}^{\rho_1} f(\delta_2)d\delta_2 \qquad (2.6)$$

2.3.5. Friction

Friction is a natural process that becomes visible only by its effects. On the one hand, this is made clear by certain forces necessary to overcome frictional resistances in any type of motion. On the other hand, friction is made manifest by tangible transition processes of mechanical energy into other forms of energy such as heat. In combination with the explanation of resistances, the term energy loss seems appropriate. But friction should not generally be associated with negative effects. In the above-given disquisition the term of friction has already been used as energy conversion in the tribological system. Following the definition of Fleischer et al. [74], friction is referred to as

> the loss of mechanical energy during the course, the initiation and the termination of a relative motion of contacting material regions.

The term of friction is not exclusively combined with surfaces or surface-near regions as the case of friction within a closed portion of matter, so-called inner friction, also complies with this definition. The processes occurring in combination with friction will be explained in more detail in the subsequent sections.

2.3.5.1. Friction types

Fleischer [74] makes clear in his tribological disquisitions that friction is always associated with the relative motion of two contacting bodies. Consequently, he uses the motion sequence in relative movement for the classification of friction types. This definition is applied for the subsequent considerations. Firstly, it is noted, however, that these types of friction in most cases may occur superimposed. Since the effects of friction very much depend on the prevailing friction type, it is necessary to clarify this type for each tribosystem to be examined.

Sliding friction Sliding friction refers to the friction type, which occurs when a pure sliding movement between the bodies is present in a tribological contact. In rotational motion this friction type is present with a slip ratio of 100 %. So pure sliding friction occurs if the contacting surfaces are purely tangentially displaced as depicted in figure 2.10.

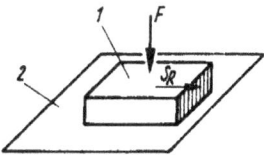

Figure 2.10.: Model of sliding friction taken from [74]

Rolling friction In rolling friction a slip-free rolling of the surfaces takes place. In this friction type at least one of the frictional bodies rotates around an axis parallel to the contact surface. The rolling elements do not necessarily have to be round shaped, as long as a cycloidal rolling takes place. The cylinder illustrated in figure 2.11, which rolls over a straight plane, represents a simple example of rolling friction. But also perfect involutes theoretically roll without any slipping on their evolutes.

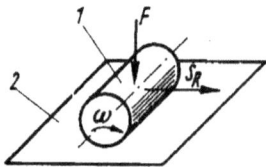

Figure 2.11.: Model of rolling friction taken from [74]

Drilling friction Also in drilling friction one of the contacting bodies executes a rotational motion. In contrast to rolling friction the rotating body rotates about an axis normal to the plane of contact as shown in figure 2.12. In other words, drilling friction can be interpreted as cycloidal sliding friction.

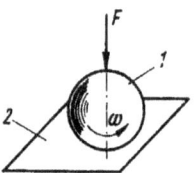

Figure 2.12.: Model of drilling friction taken from [74]

2.3.5.2. Friction states

Since the contacting material regions may exist in different aggregate states, friction may be defined on the basis of the aggregate state of the material most significantly involved in the friction process [74]. This definition is also applied in the subsequent considerations. As with the types of friction, the system response is also highly dependent on the prevailing friction state. For this reason it is necessary to clarify the friction state for any tribosystem to be examined. With the effects of the friction states, however, this clarification is by far not as trivial as with the friction types.

Solid state friction Solid state friction prevails in a tribological system, if the material region most significantly involved in friction shows solid state properties [74]. The term of dry friction, as often as it is used, does not fully represent this friction state because it implies the absence of a fluid lubricant. This, however, does not necessarily need to be the case as will be explained later on. Since in the tribological system at least two friction elements with mostly solid state properties are considered in most cases, it is obvious that these can only come into physical contact through their bounding surfaces. Surfaces of solid bodies, as explained in 2.3.4, are never absolutely smooth in the micro scale. Even if a surface is smoothened by technical means there are still asperities found, which on the micro scale will come into contact when two different surfaces rub against each other. In the solid state of bodies friction occurs exclusively in or in close proximity to solid surfaces. Therefore, all physical and chemical mechanisms that come into action when solid surfaces interact under the influence of external forces and relative motions are of particular importance in this context. The contact situation of two rubbing surfaces may very accurately be modeled as depicted in figure 2.13. Friction is directly correlated to

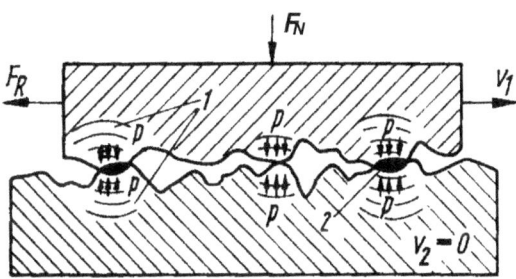

Figure 2.13.: Model of solid state friction as taken from [74] with $v_{1/2}$ = velocity of body 1/2; F_R =friction force; F_N = normal force; p = contact pressure

deformation processes occurring in the surface regions of contacting materials in relative motion. When the dynamic solid body depicted in figure 2.13 slides over the static body contacting asperities are being deformed constantly. Depending on the applied forces, the degree of deformation and the material strength of the contacting asperities, this deformation may be permanent (plastic) or reversible (elastic). But in either way energy is dissipated mostly into heat and the relative motion is inhibited. Many other mechanisms occur in this process, which will be explained later.

Liquid state friction Liquid state friction prevails in a tribological system, if the material region most significantly involved in friction shows liquid state properties [74]. It seems clear that in most tribological systems a solid contact partner needs to be available upon the occurrence of fluid state friction. If two liquid state bodies came into contact, the liquid properties would result in a mixture of both fluids. Nevertheless, flow resistances as fluid friction are consistently considered as a form of internal friction of a liquid. In a lubricated contact of two solid bodies as depicted in figure 2.14 liquid state friction prevails in the lubricant film that separates the two solid surfaces. This figure clearly shows that the liquid completely separates the two solid bodies and prevents their micro asperities from contacting directly. Again, friction may be correlated with deformation. In the case of liquid friction, however, only the separating liquid yields the load of the externally induced forces by deformation. As mentioned earlier, this deformation results in internal liquid

2. State of scientific research

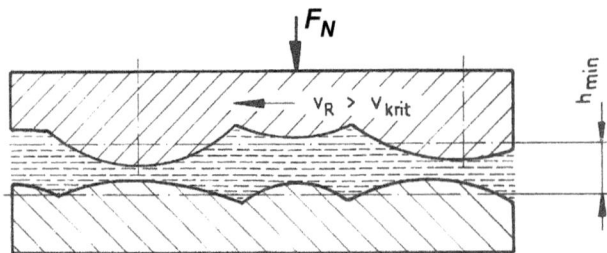

Figure 2.14.: Model of liquid state friction as taken from [24]

friction. Also this process is one of energy dissipation, which results in the generation of heat in the liquid.

Gas state friction Gas state friction is considered analogously to liquid state friction. Consequently it prevails, if the material region most significantly involved in friction shows gas state properties [74]. Statements on the mixing of gases and the internal friction apply similarly to fluid friction. Consequently figure 2.14 with all its accompanied explanation would also apply to gas state friction if the intermediate area hatched in dashed lines represented a gas instead of a liquid.

Mixed state friction The mixed friction state is clearly differentiated from all other friction states. It prevails whenever two or more friction states coexist locally or temporally [24]. The state of mixed friction is of special significance in the technical tribology. This is based on the fact that most technical tribosystems are lubricated with either liquid or partially liquid lubricants. Both, liquid and solid state friction occur in most lubricated tribosystems. Which of these states most significantly prevails in each moment highly depends on the Hertzian contact stress, the velocity of relative motion, the properties of the lubricant and the consistency of the contacting surfaces as exemplified in figure 2.15. In this figure, the friction coefficient, defined as the ratio between friction- and normal force, is plotted over the sliding speed of a pair of surfaces lubricated with a liquid lubricant. In the point of initiation, in which the relative velocity between both surfaces is still zero the friction coefficient is at its maximum value. Since there is already a separating lubricant film between the surfaces, which is stressd just like both friction bodies, this contact situation may very well be considered as mixed friction. However the ratio of solid state friction is at its maximum value in the minimum of relative velocity. The friction coefficient decreases with increasing velocity, which is attributed to the inclining share of liquid friction. Generally, liquids counter less resistance to relative motion than solid bodies. This evolution continues until a point of minimum friction is reached. This point represents a threshold to the range of pure liquid friction. Caused by hydrodynamic effects the film thickness increases in liquid friction. With further increase of velocity the resistance to relative motion also increases, which results in a growing friction coefficient. The slope of the development of friction coefficient over velocity highly

2.3. Tribology

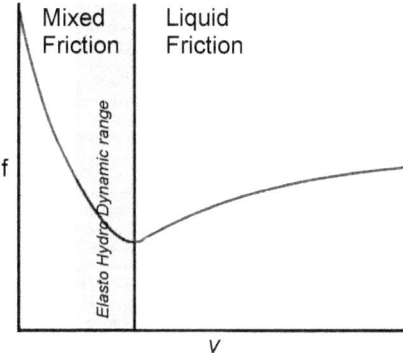

Figure 2.15.: Friction states in the Stribeck curve—taken from [24]

depends on the hydrodynamic characteristics of the separating liquid.

The state of mixed friction may very well be modeled as depicted in figure 2.16 [24]. This figure clearly shows regions in the lubricant gap separating the solid bodies

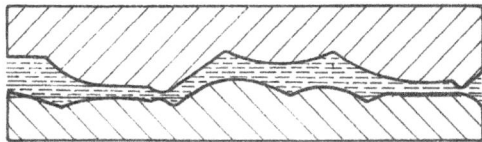

Figure 2.16.: Model of mixed state friction as taken from [24]

in which the surface asperities are completely separated without being deformed, but it also shows parts in which the lubricant separates the solid surfaces without preventing their deformation. This model contradicts the widely promulgated opinion that dry contacts may occur in well lubricated tribological systems:

> The general perception that the mixed state friction requires a tribological contact, in which there are partial areas separated by a lubricant and other partial areas in which both contacting solid bodies are in direct contact is discounted emphatically. Areas of direct contact are only possible by temperature peaks in micro regions or by a starved lubricated situation but not by a rupture of the lubricant film as a consequence of maximum pressure. [24]

Furthermore, a rupture of the lubricant film would theoretically require infinitely high pressure [24].

2.3.5.3. Friction mechanisms

The delimitation of the single friction mechanisms from the mechanisms of wear is nontrivial. In this context, evidence is found to show how closely friction and wear are

2. State of scientific research

interlinked. In the differentiation, it is important to correctly consider the cause-and-effect chain (in tribology, friction causes wear). Since wear is defined as irreversibility produced by friction it may be concluded that wear-less tribological processes are technically impossible [18, 22]. According to Czichos [75], friction mechanisms cause the energy transformation in the friction process. These friction mechanisms constitute the motion inhibiting, energy dissipating elementary processes, which take place in the contacting regions of the tribosystem. In the non-lubricated case these friction mechanisms may occur exclusively in the micro contact regions of the tribosystem. The lubricant acts as transmission medium between the micro contacts. Moreover, the lubricant extends the reach of some of these elementary processes to regions, which might not be stressed without the lubricant. This disencumbers single load peaks in the contacting surfaces. Taking into consideration the previously given definition of wear it is apparent to associate these energy dissipating elementary processes to wear. Consequently, these single friction mechanisms are linked to wear and will be discussed later on.

2.3.5.4. Friction energy

Since friction had been defined in the aforementioned way, as loss, or better, a transformation of mechanical energy in the tribological process, the term of friction needs to be expounded energetically. The fundamental law for such conversion processes is defined by the first law of thermodynamics, which states that the change of internal energy equals the sum of performed work and the procured amount of heat as depicted in equation 2.7.

$$\Delta U = W + Q \qquad (2.7)$$

By implication, this means that if work is introduced to a tribological system it can only be held in equilibrium of the internal energy if the amount of heat absorbed from the system balances the amount of introduced work. This absorption of heat is always preceded by transformational processes which will be discussed later on. All these transformational processes establish an equilibrium of the tribological frictional energy and the introduced work. With various yet to be explained friction mechanisms the friction energy can be divided into different shares or proportions. In this division, energy is differentiated into a proportion caused by deformation and another proportion caused by adhesion as displayed in figure 2.17 [74]. This chart reflects the so called dual nature of friction. With progressing subdivision, the single kinds of deformation are differentiated. These are elastic, plastic, rheological and cutting deformation. It is also differentiated between the friction bodies as each of them may uniquely react in the friction process.

Another figure which highlights the participation of each friction body in the friction process is displayed in figure 2.18. This original figure is only displayed in parts within this work but it follows this consideration with respect to different friction bodies. Grammatically this division is performed by the index of the single energy proportions. This figure depicts the energy share depending on the friction state. Moreover, this figure depicts that in pure solid state friction the energy share of both friction bodies is maximized, whilst in pure liquid friction it is minimized. In

2.3. Tribology

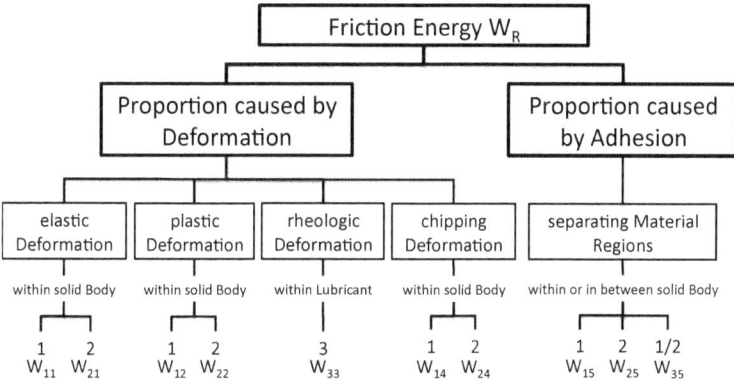

Figure 2.17.: Possible proportions of friction energy states W_{ij} [79] with i = friction body and j = mechanism—so called dual nature of friction

all forms of mixed friction the friction energy is distributed to all bodies participating in the friction process according to the currently applicable laws of distribution. In this context the special role of lubricants with respect to relief of solid surfaces should be highlighted. This consideration, although only displayed for the macroscopic

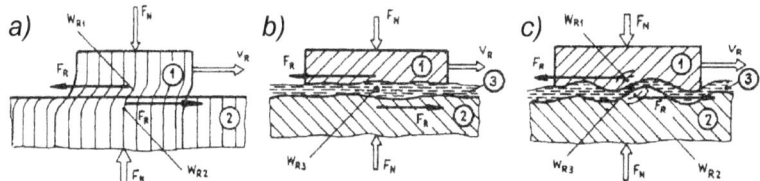

Figure 2.18.: Energetic distribution and contribution of friction bodies to the tribological process [80] — with a) Solid state friction, b)Liquid state friction, c)Mixed friction

level in figure 2.18, may be simply applied to single micro contacting spots. As this figure depicts large differences in the energy distribution depending on the friction state, the energetic approach should be subdivided for each friction state.

Calculation of solid state friction energy The most fundamental formulation of the energetic macro process of friction is signified by the work of friction as integration of the friction force over the distance traveled as expressed by

$$W_R = \int_{s_R} F_R \cdot ds_R \tag{2.8}$$

31

2. State of scientific research

The energetic approach to micro contact friction is performed by relating the sum of transformed friction energy to the volume of the involved contact. This relation results in an energy density as shown in equation 2.9 [24].

$$e = \frac{W}{V} \quad (2.9)$$

If this approach is extended to macro contacting spots or even contacting areas and if additionally the friction mechanisms according to figure 2.17 are considered, then friction energy my be quantified by equation 2.10 [24].

$$W_R = \sum_{ij} (e_{ij} \cdot V_{R_{ij}} \cdot n_{ij}) \quad (2.10)$$

In this equation n represents the number of similar energy components which prevail in the considered micro contact region and j and i refer to the friction bodies and the mechanism of deformation, respectively [24].

Relating the proportions of friction energy $W_{R,i}$ to the total sum of friction energy results in the so called energy proportion factor α_i, which distributes the friction energy to the bodies partaking in the friction process [24].

$$\alpha_i = \frac{W_{R,i}}{W_R} \quad (2.11)$$

The proportions of the single friction mechanisms have to be used for the determination of energy densities. In these disquisitions, suggestions are made for the microscopic as well as for macroscopic determination of the respective energy densities for the plastic deformation [81], the elastic deformation [82], the micro ploughing [83], the cutting deformation [84], for the separation of inter material regions [82] and for plastic impulse deformation or grooving [85]. These approaches are not repeated in this work.

Calculation of liquid state friction energy The liquid state friction is best simulated by rheometric transient shear measurements. In it, the most fundamental formulation of the energetic macro process of liquid state friction is signified by the work of shear stress, W_S, as integration of the shear stress momentum, M_S, over the angle rotated, Φ_S, as expressed by

$$W_S = \int M_S \cdot d_{\phi_S} \quad (2.12)$$

Kuhn [24] elaborated the mechanisms that occur in the shear process with references to Czarny [86] and others and explains that transient shear processes evolve in a steady state of equilibrium between the intensities of degradation of bonding structures and the reforming of the same. The energy densities that evolve from such processes will be expounded in section 2.5.1.

2.3.6. Wear

Although the term of wear has already been used in the above segments, it still needs to be defined and expounded. According to Kuhn [24], wear is defined as

> a production of irreversibility which occurs due to acting friction energy, and it encompasses all elements of a tribological system.

This definition differs from those given by other authors because it contains the term of irreversibility and because it explicitly involves all parts of a tribological system. This inclusion of all tribological system parts is displayed in 2.19. For

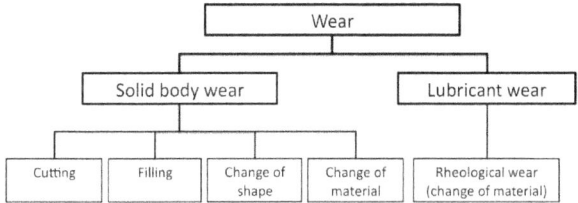

Figure 2.19.: Classification of tribological wear following [24]

further explanation and evaluation of test results only this definition is considered. The irreversible energy dissipation given in this definition completely and partially reflects the mentioned elementary processes of friction [75] because in all of these elementary processes energy is irreversibly transformed. These elementary processes are energy induction, energy transformation and energy dissipation.

2.3.6.1. Wear types

In the evaluation of wear, different types depending on the type of motion are found in analogy to the evaluation of friction. These are sliding, rolling and drilling wear.

2.3.6.2. Wear states

Wear states are defined analogously to friction states. They are categorized according to the physical state of the material most significantly involved in wear. So solid state wear refers exclusively to wear occurring in solid state material regions. The wear conditions for liquid and mixed states are interpreted accordingly.

2.3.6.3. Wear mechanisms

Adhesion Adhesion between single micro contacts with its accompanying shear and abscission of these connection points is one of the elementary processes of wear. Adhesion works by various means. On the one hand, contacting regions mechanically weld together in smallest contacting points. On the other hand physical bonds are created by the interfacial energy evoked by secondary valence bonds between surfaces. The latter may only be activated by very close approach of micro contacting regions in the friction process. Energy expenditure is necessary for the separation of any adhesively connected contacts, which inhibits the relative motion of the contacting partners [75]. Usually the softer surface yields the force so that particles adhere to the harder friction element. Lubricants in lubricated tribological systems are supposed to prevent this spot welding effect. This function is best fulfilled by a sufficient film thickess. But also distinct additives may protect surfaces from adhesive wear.

Elastic hysteresis Elastic hysteresis is initiated by elastic deformation, which takes place in the friction contact of micro asperities of distinct surface regions during the friction process. If the contacting force results in a deformation just beneath the yield strength of the deformed material it is considered as pure elastic deformation. Repeated stochastically distributed elastic deformation of this distinct asperity results in local hardening [75]. This hardening is physically caused by creeping lattice defects. Since this creeping constitutes an irreversible process it is to be regarded as wear.

Plastic deformation If smallest contacting regions of the tribosystem are deformed beyond the measure of yield strength, the corresponding solid regions start to flow and to be deformed plastically. This plastic deformation requires energy which is irreversibly transformed within the friction bodies. This process leads to an increase of the energy levels of both friction bodies.

Grooving Micro asperities harder than their respective counterparts in the friction process in combination with the acting normal load evoke grooving in the softer friction body. These asperities penetrate the softer surface overcoming a deformational resistance. This process dissipates energy. This deformation as a general rule constitutes of the single elementary processes of plastic and partial elastic deformation. The plastic deformation forms the basis of the permanent change of shape of the solid body.

Surface delamination Surface delamination occurs when the stressed surface is not able to compensate the progressive plastic deformation by pure flowing anymore. This results in cracks along the crystal lattice planes in the regions of highest stress, usually transversed to the direction of movement. The progress of this typical reaction as a surface fatigue process depends on the severeness and the number of load cycles. This fatigue process may be subdivided into an incubation period with its accompanying accumulation of lattice distortions and defects, the creation of sub micro cracks and micro cracks, the crack propagation until the unification of cracks and the terminal fracture. Single models for the creation and the propagation of cracks are discussed as a function of the brittleness of the material and the crack length [75]. The local hardening and embrittlement of the material of individual surface regions by cold deformation certainly plays a role in the mechanism of surface delamination. This points to pitting—a special case of surface delamination, in which the hardening is not preceded by very large but rather small areas of plastic deformation. The thus-hardened areas consequently are still embedded in a relatively soft base material matrix. With continuous elastic and plastic deformation of the basic matrix these hardened areas are expelled from the surface in the form of pittings.

Abrasion Abrasive wear represents the most aggressive wear mechanism. If it occurs the worn surface reveals large defects. In addition, this wear mechanism often occurs together with the other wear mechanisms, which may yet increase its aggressiveness. In the strictest sense of the term, abrasive wear is a concentration of all aforementioned wear mechanisms. It is favored by a hard/soft material combination with the softer tribological element undergoing wear in the contact. Often this wear mechanism is also favored by hard particles, which penetrate into the soft surface. In most cases, this wear mechanism is connected with very thin lubricant film thickness and starved lubrication. According to Zum Gahr [87], the abrasion process is evoked by the four subsequent subprocesses which are micro plowing, micro chipping, micro fatiguing and micro breaking as depicted in figure2.20. Micro plowing represents the most severe plastic deformation, while fatiguing is the strongest form of surface delamination. According to Czichos [75], these are initiated when a critical load of the respective working direction is reached.

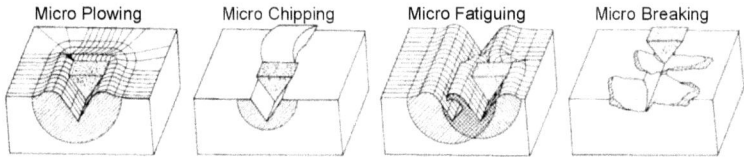

Figure 2.20.: Wear mechanisms according to [87] taken from [75]

Tirbochemical reaction In tribochemical reactions both friction bodies undergo chemical reactions, which are favored by the relative motion. These reactions may also occur in connection with the surrounding medium or the lubricant. Because of the steady erosion and deformation in the relative motion new reaction products are continually brought to the surface and the deformation provides the chemical reactions with increased target areas. Tribochemical reactions are favored by mechanical activation and the consequent creation of surface atoms with free valences and lattice defects. Moreover, these reactions are accelerated by thermal activation caused by elevated temperatures in the friction process. Reaction products may come forth as oxide regions, which are substantially harder and more brittle than the base material. Therefore they are easily removed as pittings. Tribochemical reactions are provoked easily if reaction inhibiting coatings are removed in the tribological process [75].

2.3.6.4. The energetic approach to solid state wear

In the energetic approach to solid state wear examination [24] friction energy is interpreted as the cause of wear effects. The so called apparent friction energy

2. State of scientific research

density (equation 2.13) according to Fleischer et al. [74], is defined as the link between these two concluding that

> tribological systems can only respond to frictional energy with irreversible changes [24].

Kuhn [24] also states the first law of thermodynamics in this context, upon which all energetic interpretations are based. If friction is interpreted as the total sum of consumed energy in the friction process and wear is embodied by the complete amount of worn volume, then the apparent friction energy density is formed by the ratio of these two as shown in equation 2.13 [74].

$$e_R^* = \frac{W_R}{V_V} \tag{2.13}$$

The total worn volume needs to be determined before applying this equation in the evaluation of tribological tests, which superficially seems trivial in many cases. At closer examination, however, the biggest difficulty of wear investigations becomes clear: As wear has previously been defined as the production of irreversibility, the volume of this irreversibility can not be quantified exactly. If, however, the wear volume V_V is clearly, or almost clearly, determinable, then the performed work of friction W_R may easily be referred to the wear volume. The performed friction work can be defined without great effort from the measured friction force F_R and the travelled distance s_R as shown in equation 2.14 [24].

$$W_R = F_R \cdot s_R \tag{2.14}$$

By the application of the friction shear stress τ, which is determined from the friction coefficient μ and the nominal contact pressure p_a,

$$\tau = \mu \cdot p_a \tag{2.15}$$

results the linear wear intensity I as given in the fundamental equation of wear introduced by Fleischer et al. [74]:

$$I = \frac{\tau_a}{e_R^*} \tag{2.16}$$

Application of equation 2.11 helps to distribute friction energy to the respective friction bodies as shown in

$$\frac{1}{e_R^*} = \frac{\alpha_1}{e_{R_1}} + \frac{\alpha_2}{e_{R_2}} \tag{2.17}$$

If the applied lubricant is also considered a friction body, which should happen in a consequent discussion, then equation 2.17 should be extended to

$$\frac{1}{e_R^*} = \frac{\alpha_1}{e_{R_1}} + \frac{\alpha_2}{e_{R_2}} + \frac{\alpha_3}{e_{R_3}} \tag{2.18}$$

2.3. Tribology

Apparent frictional energy density As previously expounded [24, 74], the apparent friction energy density is based on the so-called hypothesis of energy storage. This hypothesis underlies the concept that a portion of each impulse of energy is irreversibly stored in the tribologically loaded material [74]. This process of storing energy is explained by an increase of the internal energy, which is stored in the material lattice for the formation of which energy is absorbed in form of crystallization heat. If energy is introduced to the crystal lattice and this energy level is increased to the level of sublimation energy then the lattice is destroyed. The energy level necessary for the formation of a wear particle is highly decreased by lattice defects of any dimension. In the process of friction, the energy of a frictional impulse is partly stored and partly dissipated [74]:

$$W_{R_e} = W_{Store_e} + W_{Diss_e} \tag{2.19}$$

Kuhn [24] abridges Fleischer's considerations [74] by stating that friction energy is composed of simultaneous and sequential energy impulses. In addition, an energy accumulation number ζ_R, is introduced [74], which results in an expression of the proportion that is partly stored:

$$W_{Store_e} = \zeta_R \cdot W_{R_e} \tag{2.20}$$

These authors define the critical energy level necessary for the formation of wear particles by the averaged fracture energy density $\bar{e}_B$ with the application of the critical number of contacts n_k as

$$\bar{e}_B = \zeta_R \cdot e_{Re}(n_k - 1) + e_{Re} \tag{2.21}$$

Kuhn [24] further concludes that the frictional volume V_V and the wear volume V_R are interlinked by the wear number ν_V as follows

$$\nu_V = \frac{V_V}{V_R} \tag{2.22}$$

and states that the wear volume is quantified by the number of contacts and the volume of the deformed contact. With

$$\sum_{(n)} W_{Re} = n_k \cdot W_{Re} \tag{2.23}$$

and the previously given definition of the apparent frictional energy density he concludes that

$$e_R^* = e_{Re} \frac{n_k}{\nu_V} \tag{2.24}$$

Kuhn [24] and Fleischer [74] also give typical values of the energy accumulation number ζ_R and try to enumerate the number of critical contacts by the application of the Wöhler-curve and other approaches of cumulative stochastic processes. Neither of these approaches is repeated in this work.

2.3.6.5. The energetic approach to liquid state wear

The energetic approach to liquid state wear of lubricating greases is performed analogously to the energetic approach of solid state wear. So in other words, equation 2.13 would fundamentally be applicable even in investigation of liquid wear. This becomes apparent by the application of the energy accumulation and energy dissipation concepts. Although rheological wear has been defined as

> an irreversible property change of loaded material regions of a pseudo plastic lubricant evoked by a tribological load [24]

and the existence of this phenomenon has been proven by AFM investigations of stressed and unstressed grease samples [88], this approach, just like the approach for solid state wear still faces difficulties in the detection of the wear volume.

The difference between the wear approach of solid and liquid state substances is found in the gradual time dependence of wear processes within the lubricating grease, as will be shown later on in the fundamental description of rheometric evaluation approaches to lubricant wear. In other words, the complete volume of shear loaded grease will not simultaneously wear off all at once. But rather some parts of the grease are submitted to wear in the beginning of the load process while other parts are not yet involved with it. From this it is concluded [24] that the load time plays a substantial role in the wear process. Kuhn construes that all infinitesimal loaded grease parts are situated on different levels of energy accumulation that will all reach discrete individual critical load times. Furthermore, he illustrates that even in liquid wear processes friction energy is partially accumulated and dissipated. Similar to the approach of solid state wear this is expressed by

$$W_R = W_{Acc} + W_{Diss} \tag{2.25}$$

He suggests for the deformation part

$$W_{Def} = \dot{e}_{Def} \cdot V_{Def} \cdot t_B \tag{2.26}$$

where $t_B [s]$ is the load time, $V_{Def} [m^3]$ is the deformed volume and $\dot{e}_{Def} [J/(m^3 s)]$ is the applied deformation energy density per time unit. Applied to the accumulated part this leads to

$$W_{Acc} = \dot{e}_{Def} \cdot V_{Def} \cdot t_B \cdot \kappa \tag{2.27}$$

where $\kappa [-]$ is an introduced energy accumulation factor. He further concludes that a critical energy level, e_z, is reached after a critical load time, t_z.

$$e_z = \dot{e}_{Def} \cdot \kappa \cdot t_z \tag{2.28}$$

He finally concludes that these equations along with the priorly introduced relation between load time and critical load time, θ, will lead to the apparent rheological energy density

$$e*_{Rheo} = \frac{e_z \cdot \theta}{\kappa \cdot \nu} \tag{2.29}$$

where ν is the relation between frictional volume and wear volume.

2.3. Tribology

Since liquid state wear is best simulated by rheological investigations its technically implementable energetic approach will be further expounded in the fundamentals of rheometry in section 2.5.

2.3.6.6. The energetic approach to solid and liquid state wear by means of entropy balance

The above-expounded insights make clear that friction processes are always linked with energy expenditure. This energy transformation, however, inevitably denotes an irreversible process. As irreversibility, by definition, has been linked with the term of wear, it also defines the cause-and-effect relation between friction and wear. With other words, friction is the cause of wear. As all parts of a tribological system are subjected to friction they all undergo wear. This is shown by the relation

$$Wear_{system} = Wear_{solid} + Wear_{liquid} \qquad (2.30)$$

Kuhn further expounds that the process of wear as structural degradation might also be understood as a process of energy accumulation, energy dissipation and energy transition [89]. Meaning the reaching of a critical energy level. This does not only pertain to the solid but also the liquid components of the tribological system. The special focus of this work lies on lubricating greases as part of the tribological system that just like all other parts of the system partakes in wear processes. Greases as colloidal dispersed systems of base oils and thickeners undergo irreversible changes of geometry and distribution of the thickener structure as a consequence of friction energy.

Figure 2.9, already used for the explanation of the contact model of solid bodies with intermediate grease film, helps to elucidate the microsystem of a grease filled friction gap.

Because of the irreversible process characteristics of friction other authors use similar approaches for the investigation of friction and wear. Either by the attempt to find regularities in the relation between friction, heat and wear [90] or by the relation between entropy production (or entropy flow) and mass loss [90–93].

Kuhn regards the tribological system in the micro gap as an open thermodynamic system, the entropy change dS of which consists of two general terms dS_{Irrev} and dS_{Flow} [89].

$$dS = dS_{Irrev} + dS_{Flow} \qquad (2.31)$$

In this equation, dS_{Irrev} describes all irreversible mechanisms that lead to an entropy production within the system, whereas dS_{Flow} describes the portion of entropy that leaves the system by heat flow. This equation may be expressed in its differential form as

$$\frac{dS}{dt} = \frac{dS_{prod}}{dt} - \frac{dS_{Q_{1-2}}}{dt} + s_e \cdot \dot{m}_e - s_a \cdot \dot{m}_a \qquad (2.32)$$

where dS_{prod} denotes the total entropy production in the inward of the system. The term $\pm dS_{Q_{1-2}}$ leads to a change of entropy by heat transfer beyond the system

2. State of scientific research

boundaries, $s_e \cdot \dot{m}_e$ describes the entropy transport into the system by mass flow and $s_a \cdot \dot{m}_a$ denotes the portion of entropy that leaves the system by mass loss.

Each of the aforementioned inner process steps exerts its own contribution to the entropy balance. Thus it can be written with the portion of energy accumulation, S_{acc}, the portion of energy dissipation, S_{diss} and the portion of energy transition, S_{trans}

$$\frac{dS_{prod}}{dt} = \dot{S}_{acc} + \dot{S}_{diss} + \dot{S}_{trans} \tag{2.33}$$

Inserted into equation 2.32 it results in

$$\frac{dS}{dt} = \left(\frac{dS_{acc}}{dt} + \frac{dS_{diss}}{dt} + \frac{dS_{trans}}{dt}\right) - \frac{dS_{Q_{1-2}}}{dt} + s_e \cdot \dot{m}_e - s_a \cdot \dot{m}_a \tag{2.34}$$

Moreover, Kuhn suggests the following relation for S_{acc} containing the energy density necessary for the deformation process e_{def}, the portion of accumulated friction energy ζ_R, the accumulation volume V_{acc}, and the temperature of the accumulation process T_{acc}.

$$S_{acc} = \frac{e_{def} \cdot \zeta_R \cdot V_{acc}}{T_{acc}} \tag{2.35}$$

Additionally, Kuhn suggests for S_{Trans},

$$S_{trans} = \frac{G' \cdot \gamma_{crit}^2}{cos\delta \cdot T_{trans}} \tag{2.36}$$

where γ_{crit} is the critical shear deformation and T_{trans} is the temperature of the transition process. With these relations he models the contact situation along with the entropy transport into the system as shown in figure 2.21.

Additionally, Kuhn suggests for the relation between the apparent friction energy density e^*_{Rrheo} and the entropy density leaving the system

$$e^*_{R\ rheo} = T_f \cdot (\rho_a \cdot s_a) - \frac{T_f}{V_a} \cdot (S_e - S_{Q1-2}) \tag{2.37}$$

with the density of the mass of material leaving the system ρ_a, and the entropy density leaving the system s_a, the friction temperature T_f and the volume of the mass leaving the system V_a.

2.4. Tribometry and evaluation of tribological data

As sliding contacts in real life applications evoke highest wear rates, they are mostly tried to be prevented in mechanical elements by means of design and structure. In gears, e.g. sliding resulting from slip between the tooth flanks is prevented by involuting tooth shapes. Rolling elements in roller bearings also prevent the surfaces of the inner and outer ring to slide against each other. Tribological tests are usually performed to monitor friction and wear responses of surfaces in the lubricated contacts over time. In order to come to soon results, these tribological

2.5. Rheological tests and energetic interpretation

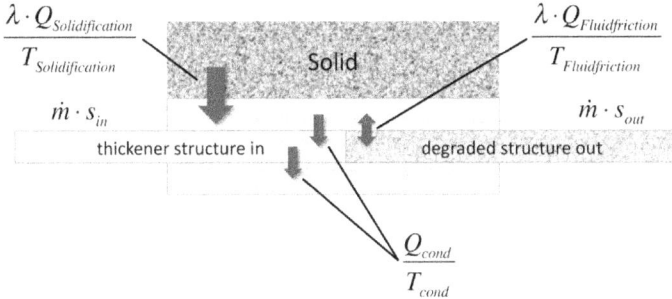

Figure 2.21.: Model of entropy transport or entropy flow into the system [89]—including the thermal conductivity λ, heat quantity of solid and fluid friction $Q_{Solidfriction}$ and $Q_{Liqidfriction}$, respectively, the temperature of solid and liquid friction $T_{Solidfriction}$ and $T_{Liquidfriction}$, respectively and the temperature and quantity of conducted heat between grease layers T_{cond} and Q_{cond}, respectively

series of experiments are often performed in sliding contact situation. This way, wear rates are intensified and the time elapsed to investigate wear characteristics is drastically reduced. Still there are many standardized tribological tests, which try to simulate real life applications in bearings and gears. Therefore, these tests forgo a pure sliding contact. In all of these tests abnormally high loads are applied to produce wear within a short period of time, which is then extrapolated to life spans of machine elements in real life applications.

2.5. Rheological tests and energetic interpretation

In the above-given disquisitions little was mentioned about the liquid state friction and wear. Since these two friction and wear states are only rheologically measurable they are consequently subsumed under the heading of rheology.

2.5.1. Rotational transient tests

Rotational transient tests are usually performed to monitor the shear stress response over time at constant shear rates in order to simulate the fluid friction state within grease-lubricated tribological contacts. The typical grease behavior in such rotational tests with constant shear rates, as previously described [11, 19, 86], shows a peak of maximum shear stress, which is related to the elastic response of the grease structure. This peak is followed by a steady decline as a result of the structural degradation of the grease and its consequent loss of viscosity. Subsequently, the shear stress responds with a convergent approximation to a limiting stable shear stress representing an equilibrium between shear-induced microstructural remaining stability and the applied shear rate [11, 19, 94]. A typical evolution of shear stress over time is plotted in figure 2.22. This figure clearly depicts the aforementioned

2. State of scientific research

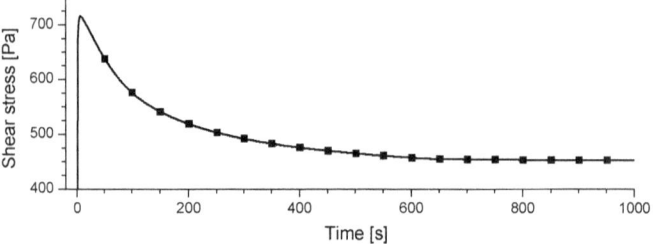

Figure 2.22.: Example curve of a rotational transient test

time-dependence structural degradation.

The energetic approach for evaluation of rotational tests

Following the methodology described by Kuhn [11], further developed a mathematical approach to describe the grease structure degradation process in transient flow tests [13, 15–17, 19, 24], the shear stress decay over time may be described as

$$\tau(t) > \tau_{max} \Rightarrow \tau(t) = \tau_{lim} \cdot \left(\frac{t}{t_{lim}}\right)^{-n} \qquad (2.38)$$

where τ_{max}[Pa] is the stress overshoot, τ_{lim}[Pa] is the remaining shear stress after completion of the degradation process, i.e. the steady-state value, t[s] is the working time, t_{lim}[s] is the working time until reaching τ_{lim} and n[-] is a dimensionless exponent for the description of the degradation intensity. In further discussion of this approach [30] the mathematical description of the elastic deformational process, leading to the maximum start value of shear stress, τ_{max}, may be neglected in this work, which focuses on the shear-induced structural degradation process. Generally, the integration of the $\tau(t)$ curve for any given period of time is to be interpreted as energy expenditure per stressed grease volume, namely the rheological energy density [24]. This is expressed by

$$e_{rheo-rot} = \dot{\gamma} \cdot \int_{0}^{\zeta} \tau(t) dt [\text{J}/\text{m}^3] \qquad (2.39)$$

where $\dot{\gamma}$[s^{-1}] is the given constant shear rate applied in the rotational test and ζ[s] is the evaluation time period—in the case of the interpretation of this work it is to be set equal to τ_{lim}. In previous studies [22], the energy density approach was used to interpret wear as a direct consequence of friction, so that the definition of wear is expanded and liberated from the limitation of the sole solid state unto the introduction of liquid state wear, meaning the degradation processes and the production of irreversibility in the separating lubricant film as a reaction to the applied frictional energy. Consequently, friction may be defined as an

2.5. Rheological tests and energetic interpretation

an irreversible process, which leads to the transformation and accumulation of mechanical energy accompanied by the production of entropy [22].

Therefore, the aforementioned degradation process in rotational tests may be expressed as rheological wear.

2.5.2. Oscillatory shear tests – Amplitude sweeps

Lubricating greases are viscoelastic substances that react in multiple ways to shear stress applied to it. As their name implies, they can viscously and/or elastically react to shear stress with lasting or reversible deformation, respectively. The elastic character delineates their ability to reversibly store the mechanical energy introduced from the exterior, whereas the viscous characteristics points to their ability to internally dissipate this portion of energy by irreversible conversion into heat. These characteristics can be differently pronounced in any type of viscoelastic material depending on the type and mode of application of external load. As both of these mentioned characteristics are inherently combined in viscoelastic substances they will react to shear stress with a delayed response, exhibiting a time dependent behavior that cannot be defined by material constants only. Their viscoelastic behavior can be described by linear differential equations with constant coefficients if deformations and stresses are considered infinitesimally small. In this case, pure time dependence prevails and material responses are regarded as linear viscoelastic. Creep, relaxation and periodic deformation tests are suitable methods to investigate such behavior [95], only the latter of which will be expounded subsequently. Small amplitude oscillatory shear (SAOS) measurements have been entrenched as one of the main examination methods for the rheological characterization of lubricating greases. This is based on the fact that greases exhibit an extremely distinctive non-Newtonian viscoelastic behavior, which by means of oscillatory measurements, can be tested more intensely than with traditional measurements of grease consistency.

If a viscoelastic substance, i.e. a lubricating grease, is loaded with an harmonically periodic, sinusoidal oscillating shear stress it will likewise respond with a sinusoidal deformation oscillation. In the regarded example of sinusoidal shear oscillation

$$\tau = \tau_0 cos(\omega t) \tag{2.40}$$

the strain will react as

$$\gamma = \gamma_0 cos(\omega t + \delta) \tag{2.41}$$

where $t[s]$ is the time, τ_0 is the maximum stress amplitude, γ_0 is the maximum deformation amplitude, $\omega[rad/s]$ is the oscillation circular frequency and $\delta[-]$ is the phase shift angle.

This equation clearly shows that the response oscillation will exhibit the same frequency with a phase shift angle of δ. This phase difference can result within the limits of $0 < \delta < \pi/2$. It indicates the behavior of the tested sample in each respective measuring point, where $\delta = 0$ points to complete elastic behavior and $\delta = \pi/2$ points to absolutely viscous (liquid) behavior as highlighted in figure 2.23.

The oscillation equations can also be outlined in their complex forms [96]

$$\tau(\omega t) = \tau_0 \cdot e^{i\omega t} = \tau_0(cos(\omega t) + i \cdot sin(\omega t)) = \tau'(\omega) + i \cdot \tau''(\omega) \tag{2.42}$$

2. State of scientific research

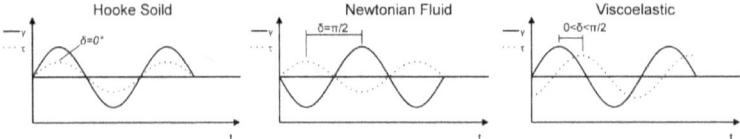

Figure 2.23.: Example curves of characteristic sinusoidal load in small amplitude oscillatory shear (SAOS)

$$\gamma(\omega t) = \gamma_0 \cdot e^{i\omega t + \delta} = \gamma_0(cos(\omega t + \delta) + i \cdot sin(\omega t + \delta)) = \gamma'(\omega) + j \cdot \gamma''(\omega) \quad (2.43)$$

where the real parts are represented by (') and the imaginary parts are represented by (''). The complex shear modulus, $G*$, the main parameter for the definition of flow characteristics in periodic oscillation, is defined by [95, 96] as

$$G^* = \frac{\tau_0}{\gamma_0} = \frac{\tau_0}{\gamma_0}(cos\delta + i \cdot sin\delta) = G' + iG'' \quad (2.44)$$

The elastically responding share of the measured substance is represented by G' as a component that evolves in phase with the shear stress. It is defined as a measure of the deformation energy that is elastically stored in and released from the material during the shearing process [97]. For this reason it is referred to as the storage modulus. G' and G^* are linked via the cosine of the loss angle, δ.

$$G' = G^* \cdot cos(\delta)[Pa] \quad (2.45)$$

The viscous share of the material, on the other hand, is represented by G''. It is defined as a measure of the kinetic energy that is dissipated during the plastic flow. Consequently it is referred to as the loss modulus and it is linked to the G^* via the sine of the loss angle δ.

$$G'' = G^* \cdot sin(\delta)[Pa] \quad (2.46)$$

If determined values of G' are larger than values of G'' in a certain stress (or strain) range, solid body properties prevail over liquid properties. As long as $G'' > 0$ one would refer to this substance as a viscoelastic solid body. If, in reverse, a certain measuring point resulted in G'' values larger than those of G' the substance would be referred to as a viscoelastic liquid [97].

The relation of these two is defined as the the loss tangent, $tan\delta$

$$tan\delta = \frac{G''}{G'} \quad (2.47)$$

as such, it completely disregards the absolute values of these moduli but rather defines a relative magnitude that is able to assert which of the elastic or plastic deformational state prevails in each measuring point. Values of $tan(\delta) > 1$ point to a predominating plastic characteristic of the measured substance in that specific measuring point, whereas values of $tan(\delta) < 1$ point to viscoeslatic characteristics predominated by elastic behavior. In other words, within the limits of $0 < tan(\delta) < 1$ one might generally say that elastic behavior predominates but the plastic influence increases along with $tan(\delta)$.

2.5. Rheological tests and energetic interpretation

The absolute value of G^* is defined by

$$|G*| = \sqrt{G'^2 + G''^2} = \frac{\tau_0}{\gamma_0} \qquad (2.48)$$

In amplitude sweep oscillatory rheological measurements, a suitable amount of grease is measured usually with plate/plate or cone/plate systems. After setting the measuring gap, the sample is provided with enough time to relax from the strain and stress that occurred while setting the gap. Afterwards, one plate begins to oscillate rotatorily with a preset sinusoidal frequency and test duration around the axis normal to the measuring surface. The amplitude of the oscillation is steadily increased according to a preset pattern. The measurement usually starts with very small amplitudes. Depending on the rheometer setup the amplitude is either modulated by a set strain (γ [m/m]) or by a set stress (τ [Pa]). Which of these physical values (stress or strain) is selected to be the manipulated variable, depends on the discretion of the investigator. Independently of the parametrization of the measurement, both physical values (stress and strain) may be selected to be plotted on the axis of abscissa in the subsequent evaluation of the measurement. The axis of ordinates usually plots the values for G' and G'' in duplex logarithmic scale over the stress or strain as depicted in figure 2.24.

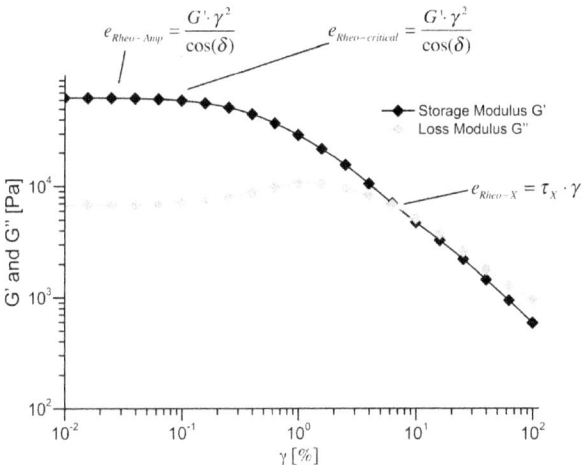

Figure 2.24.: Example curve of an amplitude sweep rheometer test including characteristic points

Typically, lubricating greases respond to amplitude sweep tests as shown in figure 2.24. In the beginning of the test, under small amplitudes, G' results in values that are some orders of magnitude larger than the ones obtained for G''. So with small amplitudes, greases behave like a solid body. This effect can be seen even when a grease container is opened and tilted—it does not flow under the influence of gravity. This means a certain force is necessary to make the grease flow. The plots of both physical quantities continue in a constant linear plateau. This plateau is called the linear-visco-elastic range (LVE). The magnitude of this range constitutes a specific characteristic of the grease. Within this range, there is a proportionality between the applied oscillatory stress and the resulting strain, or vice versa. As a result, all the linear viscoelastic functions, like G' and G'', do not depend on the magnitude of the stress (or strain) applied. Only a change of either of the two shear moduli (G' or G'') of more than 10 % from the LVE-plateau marks the limit of the LVE range according to DIN 51810-2. In this determination it does not matter if either value increases or decreases. With some measured substances the loss modulus G'' exhibits a 'distinct hump' even before G' values start to change. This points to a

> portion of deformational energy spent to irreversibly deform substructures of the substance even prior to the complete structural break-down [97].

This point initiates the partial flux of the measured substance. Since the material starts to flow, this point is defined as 'yield point' according to DIN 51810-2. Subsequent to this yield-point, G' usually progresses with a steady decline, as well as G'', often after reaching a local maximum. Since G'' results in less slope than G' they intersect at another characteristic point—the crossover point. This point marks the threshold to pure flowing of the substance. For this reason, sometimes it is also called the flow point.

The energetic approach for evaluation of amplitude sweep tests

Because the energetic term was already used for the explanation of the two physical quantities of G' and G'' it seems reasonable to also apply this approach to the interpretation of the data acquired by rheological amplitude sweep measurements. Assuming that both the tribological and rheological processes, in which friction and wear are involved as processes of energetic load, energy accumulation and energy transition at critical energy levels and consequent energetic relieve of the system [23], it is suggested to interpret the rheological load of a grease within the LVE-range as rheological friction. Just as was done in the derivation of the energy density of the solid state friction [24], Kuhn [23] sets the loss of energy in relation to the stressed volume and interprets the left hand half of the equation (F_R/A_R) as the shear stress in rheological tests τ and the right hand part (s_R/h) as the strain γ.

$$\frac{W_R}{V_V} = \frac{F_R \cdot s_R}{A_R \cdot h} = \tau \cdot \gamma [\mathrm{J/m^3}] \qquad (2.49)$$

where $A_R [m^2]$ is the frictional area and $h [m]$ is the wear hight. Inserting equations 2.44 and 2.45 into equation 2.49 results in

$$e_{Rheo-Amp} = \frac{G' \cdot \gamma}{cos(\delta)} \cdot \gamma = \frac{G' \cdot \gamma^2}{cos(\delta)} [\mathrm{J/m^3}] \qquad (2.50)$$

Therein, the whole stressed volume is not taken as a basis to the consideration but only the volume share of the thickener content of the lubricating grease. This is based on the assumption that the thickener is the only part of the grease that would be subjected to wear. Depending on the prevailing amplitude in the oscillatory

2.5. Rheological tests and energetic interpretation

test, this rheological friction can reach certain energy levels. A verification of the physical units in this equation validates the results. With J/m^3 it results in energy per volume. Moreover, different energy levels are established depending on their position in the LVE-range and beyond. Since it is deduced that a critical energy level is reached in the yield-point, which initiates the flux of the grease, this energy level is named $e_{rheo-critical}$. Analogously, the energy level in the cross-over point is labeled e_{rheo-X}. At this point, equation 2.50 can be set equivalent to

$$e^*_{Rheo-X} = \frac{W_f}{V_W} = \tau_X \cdot \gamma_X \, [J/m^3] \qquad (2.51)$$

The index X of τ and γ indicates that with this equation the cross-over point is regarded exclusively. This disquisition allows to energetically evaluate the measurement results of oscillatory rheological tests.

2.5.3. Oscillatory shear tests – Frequency sweeps

Oscillatory tests with constant amplitude and inclining frequency are another method of rheological investigation. These frequency sweep tests are performed in order to determine the time-dependent behavior of the measured substances [97]. Results at high frequencies simulate the short-time behavior of the substance by rapid movements and results at low frequencies simulate their long-term characteristics by slow movements. This way, frequency sweeps reveal how the measured material changes as a function of stress or strain application rate. As frequency sweep results leave room for interpretation of the materials' microstructure they are referred to as their rheological "finger prints" [98].

In frequency sweeps the dynamic plate of the rheometer is brought to sinusoidal oscillation with a logarithmic or linear frequency ramp. Here again the choice is left to select between the deformation γ either in [%] or [m/m] and the shear stress τ [Pa] as the actuating variable, depending on the rheometer. Traditionally, in order to obtain the mechanical spectrum of a given material, frequency sweeps are performed with amplitudes within the linear viscoelasticity-range. For this purpose amplitude sweeps need to be performed prior to the test in order to determine the LVE-range. From the experimental point of view, usually the selected amplitudes should not be right in the beginning of the LVE-range because too small amplitudes could provoke instabilities and errors in the rheometer. On the other hand the selected amplitudes should not be too close to the end of the LVE-range either, because too large amplitudes could provoke measuring points outside the LVE-range at high or low frequencies. So the selected amplitude should be at a relatively large but still safe deformation. Also, in frequency sweeps the evolution of the two shear moduli G' and G'' are plotted in a diagram in log-log scale representing the mechanical spectrum typically with values of G' around one decade higher than G'' values in the in the complete measured frequency range [28, 29, 99–101]. These same authors denote that the course of G'' in combination with lubricating greases renownedly depicts a distinct minimum as well as a power-law relationship with very little slope between G' and the frequency, which is typical for highly structured systems. This part of the mechanical spectrum as a plateau region is highly typical for entangled polymeric systems as well as gel-like dispersions.

The energetic approach for evaluation of frequency sweep tests

Evaluation of frequency sweeps according to the energetic approach works just like it does with the amplitude sweeps. Since the values of the quantities G' and G'' are monitored as a function of the frequency it seems only appropriate to take these and insert them into the known rheological energy density equation 2.50. The corresponding values of G' and $cos(\delta)$ can be averaged either for specific points, for ranges of values or the complete curve profile. Since the values of γ remain constant in the whole curve, the preset γ can simply be inserted into the equation.

2.5.4. Tensile tests

Tensile tests provide a method of grease investigation introduced by engineers to generate another method to be energetically simply interpreted. The experiment is performed on the basis of tensile strength and ductility tests for solid samples. Most rheological tests take advantage of the rotational symmetry of the measurement systems. The advantage of rotating around the main axis of the measuring system is obvious. The measuring path can be set infinitely long resulting in infinite revolutions, and kinetic energy is introduced into the substance by rotating around the main axis. In tensile tests, however, it is introduced in the normal direction. In these tensile tests a suitable amount of grease is added to the rheometer, the predefined gap is set and then, after the removal of excess volume of grease, the gap is increased with a preset velocity. The resulting normal load is measured as tensile force and calculated with the diameter and the resulting area of the measuring system into the tensile stress. It seems obvious that in such tensile tests a peak of maximum stress is reached right after the start of the test. Soon after the beginning of the test, constricting effects occur and, because of lacking elasticity, the grease film tears off pulling strings behind.

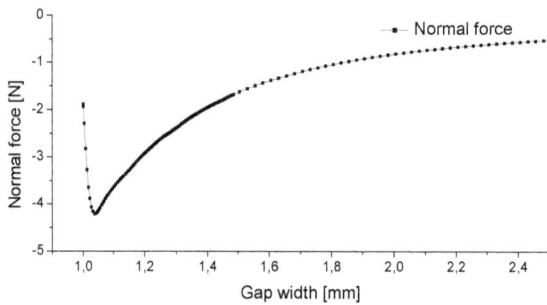

Figure 2.25.: Example curve of a tensile test

Figure 2.25 shows an example curve of a tensile test performed with preset starting gap of 1 mm. With tensile forces being defined negative, this graph exhibits a steep slope down to a negative maximum right in the beginning of the force application. After this maximum is crossed, the aforementioned constricting effects lead to a decline. As the start phase of such curves including this maximum is most interesting for evaluation, it seems plausible to set up the measuring device to collect as many measuring points as possible during this phase.

The energetic approach for evaluation of tensile tests

The energetic evaluation of rheological tensile tests is conducted in accordance with the energetic approach to tensile tests of solid material. In such tests where materials are put under the influence of defined stresses, the stress is usually plotted as a function of the strain. The area under the stress-strain curve is defined as the energy per volume put into the stressed system. In case of pure elastic behavior this energy, as potential energy, is stored in the system and is not subjected to any transformation processes. Thus it is completely, reversibly and immediately returned to the surroundings upon relief of the stress. In the case of plastic behavior, a portion of the energy put into the system is subjected to transformational processes, which result in irreversible plastic deformation, internal friction and generation of heat and noise. But still, the area under the stress-strain curve defines the total amount of energy per volume inserted into the system. In the evaluation of greases by means of tensile tests, this energy as it is related to the stresses volume should be called a rheological energy density $e_{rheo-tensile}$. With the stress $\sigma = \frac{F}{A}[\text{N/m}^2]$ and the strain $\epsilon = \frac{\delta l}{l}[\text{m/m}]$ the total energy density results from

$$e_{rheo-tensile} = \int_0^\epsilon \sigma d\epsilon [\text{J/m}^3] \tag{2.52}$$

The limits of integration are usually defined from zero elongation, meaning the unstrained substance, until the maximum elongation $\epsilon_{rupture}$, meaning the strain at the moment of tear-off. With the tensile force $F[\text{N}]$, the stressed area $A[\text{m}^2]$ and the length of elongation $l[\text{m}]$ this equation develops as follows

$$e_{rheo-tensile} = \int_0^{\epsilon_{rupture}} \sigma d\epsilon = \int_0^{l_{rupture}} \frac{F \cdot dl}{A \cdot l} = \frac{1}{V} \int_0^{l_{rupture}} F dl [\text{J/m}^3] \tag{2.53}$$

In further application of this equation, it has to be kept in mind, however, that this equation relates the energy to the total amount of applied grease.

2.6. Other tests

2.6.1. Bouncing ball tests

Bouncing ball tests define a test method, which allows to investigate how well the greases are able to protect the surface under extreme pressure conditions. Initially, these bouncing ball tests were intended to investigate the cohesion behavior of the greases. In these tests, a standard roller bearing steel ball is attached to a vacuum holder in a defined hight over a metal surface. A grease layer is applied to this metal surface. In the preparation of these tests it is very crucial to always apply the exact same amount of grease to an evenly ground and polished metal surface. Not only the amount of grease is important but especially the thickness and the homogeneity of the grease layer. The steel ball is instantaneously released from the gripper, which then accelerates due to gravity and falls until it collides with the grease coated metal surface. The gripper is designed to release the ball without any rotatory acceleration. In the moment of impact the kinetic energy accumulated from the potential energy during the downward acceleration deforms the grease layer, the metal surface and the steel ball plastically and elastically. The total amount of energy used to elastically deform any parts of the surfaces is retransmitted to the ball and accelerates it in the opposite direction of the free fall. The amount of energy spent to plastically deform the ball, the intermediate grease layer and the steel surface will not be transmitted to the ball resulting in a potential difference between the starting hight of the free fall and the maximum hight of the first ricochet. This hight difference is detected by means of a high-speed camera and can easily be calculated into an energy difference. If h_1[m] is the initial hight and h_2[m] is the maximum hight after the first rebound and m[kg] is the mass of the steel ball then this energy difference can be calculated with g[m/s^2] with

$$\Delta E = m \cdot g (h_1 - h_2) [J] \qquad (2.54)$$

Next to the hight and the corresponding energy there are also other things, which can be investigated with these kinds of tests. With the amount of energy transformed into plastic deformation of the polished steel plate one may also closely examine the intensity of the deformed part of the metal surface and the shape and diameter of the imprint. This analysis especially may allow to draw conclusions on the role of the grease as a separating medium in highly loaded tribological contacts. The dimension of the plastically deformed imprint in the polished metal surface along with its material characteristics gives information about the amount of energy transformed for plastic deformation during the impact. On the other hand the resulting hight difference after the first rebound as determined with equation 2.54 gives information on the total transformed energy. The difference between this value and the amount of energy spent for plastic deformation of the metal surface consequently should bring to light the proportion of energy spent for the plastic deformation of the intermediate grease layer, next to the presumably much smaller proportion of energy dissipated by noise and by gas friction in the falling and the rising processes. The total amount of heat generated by inner friction in the plastic deformation is already covered by the equivalent stress hypotheses in the calculation of energy needed to deform the metal surface. This way the cohesion characteristics of the grease and the energy spent for its plastic deformation could be evaluated. In order to give exact comparable information on the total energy spent to plastically deform the intermediate grease layer one could also evaluate bouncing ball tests by performing two sets of tests—one with a test setup including a grease layer and one without any grease.

Apart from the energetic evaluation of determined data in bouncing ball tests, these experiments leave room for another method of analysis pertaining to the moment following right after the impact of the ball. With applied grease layers on the impacted surfaces, the bouncing ball, after deforming the lubricant, usually drags some amount of grease in the following upward acceleration. This portion of grease forms a filament, which is extended to some extend with constricting effects before it breaks. A high-speed camera may capture this event and help to evaluate the maximum lengths of these grease filaments.

3. Assumptions and models

The working mechanisms of greases on the micro scale are exceedingly complex. Physical as well as chemical aspects influence the behavior of lubricating greases in the frictional contact. In the formulation of lubricating greases, manufacturers and developers have to rely on their experience in order to find the formula that best fits the desired application. As the main focus of all investigations in this work is directed to bio-lubricating greases, which mostly consist of esters of fatty acids derived from vegetable oils, it also covers the likelihood of highly polar base oils that naturally come with the application of esters. Base oils of bio-greases with their high polarity, more than the low polarity mineral base oils of traditional greases, influence the tribological and rheological behavior of the finished product. In traditional grease formulations based on mineral oils, polar-active working compounds are added mostly after the grease formulation to attain desired lubricant characteristics. As this controlling mechanism is highly interfered when the base oil already contains highly polar active compounds it is even more important to understand their physical working mechanisms. Assumptions regarding how base oil polarities affect the tribological and rheological behavior of the greases will be postulated to gain a deeper understanding in this matter. Therefore the influence of the molecular structure of the base oils' main components as well as tribologically active surfaces will be taken into consideration.

3.1. Assumptions and deriving models for the microphysical behavior of greases

From the current state of research, the assumption that polarities of lubricating grease base oils substantially affect the grease cohesion as well as their type and dimension of interaction with tribological surfaces can be derived. This interaction highly depends on the bonding character of participating molecules in the base oils, the thickeners and the tribological surfaces. The thickeners as isolated and dried out substances would neither cohere nor adhere to a metal surface without cooperation of the oils. Therefore, special focus will be drawn to the base oil-thickener-surface interaction. As lubrication works by the mechanisms described in the state of current research, it is assumed that, especially in the state of mixed friction, rheological as well as tribological effects cause all frictional and wear results. This assumption will be the basis of all subsequent postulations and experimental work as well as theoretical explanations. The postulations to be proved subsequently are as follows:

- The substantial influence of oil polarity on the interaction of greases with tribological surfaces affects the tribological behavior of lubricating greases. This will show in the frictional and wear behavior.

3. Assumptions and models

- The substantial influence of oil polarity on the cohesion of lubricating greases affects their rheological behavior.
- The influence of oil polarity on the rheological behavior of greases will also affect their tribological characteristics.

Based on these postulations three distinct models can be derived, which will help to elucidate the effects to be investigated.

3.1.1. Model A – physical working mechanisms of oil polarity affecting the tribological behavior

The tribological model taken from [51] describes the tribological activity of certain additive packages, which physically interact with the tribological surface. In the disquisition of this work this model is applied to the bonding type of the base oil molecules. Physisorptive interchange of base oil molecules and tribosurfaces rests upon van-der-Waals interactions. Their tendency to occur is explained by

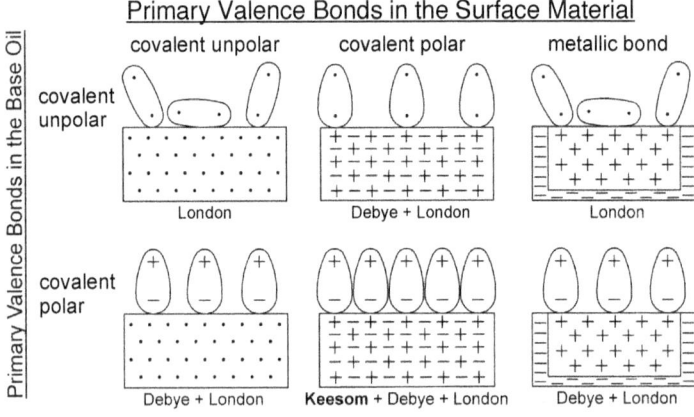

Figure 3.1.: Model of oil polarity influence on tribological response [51]

the bonding type of participating oil and surface molecules as depicted in figure 3.1. This figure represents a table of images with the base oil polarity changing in the rows and the surface polarity alternating in columns. The oil molecules are represented by ovals the sign, notation and shape of which delineating their polarity. The solid surfaces are represented by blocks. The surface polarity is also represented by internal notation and signs. If for example base oils consisting of mainly covalent unpolar bonds come into contact with an unpolar surface such as sapphire, then only van-der-Waals interactions of London type may occur. The weakness of this physisorption is modeled by the misalignment of oil molecules on the surface as depicted in the top left image in figure 3.1. All other types of van-der-Waals interactions resulting from different combinations of primary valence bonds in the oils as well as the surfaces are displayed in this figure. It should be

3.1. Assumptions and deriving models for the microphysical behavior of greases

highlighted that Keesom forces are only evoked by a polar-polar combination. Also it should be mentioned that this model needs to be applied to both friction partners in the tribological contact. Therefore, if two surfaces of different polarities contact each other then this model should be applied to both surfaces resulting in different protection mechanisms for each individual surface.

3.1.2. Model B – physical working mechanisms of oil polarity affecting the rheological behavior

As polarity differences between base oils affect the tribological behavior of the actively participating and wetted surfaces in the contact by mentioned influences it is expected that they also exert a substantial influence on the thickener structure. Different types of thickeners, as described in the state of scientific research, work by individual mechanisms. The base oil polarity will exert a substantial influence on the thickening behavior as all these mechanisms in their own individual manner establish bond forces between the oil and the thickener structure. This will influence the

- absolute percentage of thickener necessary for greases in equal consistency grade formulated with base oils of different individual polarity
- the rheological in transient shear
- the rheological in small amplitude oscillatory shear (SAOS)

3.1.3. Model C – interdependance between rheological and tribological characteristics of greases influenced by base oil polarity

As lubrication in the mixed friction state highly involves the base oil interactions with the surfaces as well as the rheological characteristics of greases, it is concluded that base oil polarities exert a twofold influence on the frictional and wear results—base oil polarities influence the surface interaction as outlined in model A as well as the rheological characteristics as outlined in model B. For this reason, model C interlinks the mutual influences of models A and B. Which of these respective models have a larger effect on the tribological response highly depends on the characteristics of each present tribosystem. There might be tribosystems, in which model A prevails over model B and vice versa. The subsequent investigations will have to bring to light both extremes in order to validate these models.

4. Materials, methods and conditions for the experimental analysis of greases

4.1. Selection of relevant materials

The assumptions and models and the mechanisms affecting the tribological response of biodegradable greases studied need to be experimentally validated. For this reason the tribological behavior has to be investigated by means of suitable series of tribological experiments, in which frictional and wear responses of appropriate materials will be evaluated. These materials have to be selected to cover a wide spectrum of different polarities. Also specific methods capable to disclose the respective effects have to be selected. In addition, the rheological behavior of the sample greases yet to be selected has to be investigated by relevant tests in order to prove the different postulated models. Test parameters and conditions of this rheological investigation have to be selected in a way capable to give evidence to the model.

4.1.1. The greases

Qualified greases need to be selected to prove the assumptions made in section 3.1. In this background it is of prime importance to select raw materials, which cover a wide range of different oil polarities, thickener types and biodegradability or toxicity degrees. This way, research is performed on a wide basis. On the other hand it is also clear that the number of different raw materials should not be too large as this would largely increase the number of greases and consequently the number of tests to be performed. While a large number of different greases would limit the number of performable tests down to a few, a small number of greases would lead to a large test variety. The number of greases and the number of test to be performed was well balanced for the research performed in the present work. This led to a limitation of the number of greases to sixteen in total. The matrix of grease candidates to be evaluated was decided to be evenly distributed by four different types of thickeners with four different base oils. It was decided to formulate all sixteen greases without any lubricant additives to be able to attribute potential effects to be found solely to the thickener system or the applied base oils. For the same reason it was decided to formulate all sixteen greases to the same consistency grade, which was decided to be NLGI grade 2 according to DIN 51818 and ISO 2137—the most common consistency grade for grease applied in roller bearings.

4.1.1.1. The base oils

Oils of different polarities are needed as base stock for the grease investigations necessary to prove the assumptions postulated in section 3.1. In order to better differentiate the investigated effects pertaining to base oil polarity it was decided to chose two high-polarity and two low-polarity base oils. Since it is known that biogenic oleochemicals, especially esters of long chain fatty acids, exhibit high oil polarities two such candidates were chosen as base oils. To enable a wide variety of differently polar base oils in the formulated greases it is necessary to make sure that low polarity oils are also applied in the grease matrix. The following oils, mainly esters, were chosen as base oils for the grease candidates to be formulated:

- *High-oleic sunflower oil* (HOSO), an ester with a ratio of over 90 % of oleic acid (C18:1) and a trivalent glycerol as the alcoholic component. This ester is of 100 % regrowing origin. It is gained by a special breed of sunflowers with an extra high share of oleic acid in the acid component of the ester. Other fatty acids such as palmitic, stearic and linoleic acids are also found in small concentrations within HOSO. Because of its trivalent alcohol with three C=O double bonds, HOSO is suspected to have a very high polarity.

- *trimethylol propane trioleate* (TMPO) is another high-polarity base oil candidate. This ester comprises of a synthetic trimethylol alcohol with three oleic acid (C18:1) branchesThe acid component of this ester is gained from rapeseed. Because of the synthetic origin of the alcohol component this ester has a share of regrowing ingredients of only 80 %.

- *12-octyl dodecyl isostearate* (OCT) is a synthetic guerbet ester, which consists of a monovalent alcohol and stearic acid (C18:0). All its components (alcohol and acid) are based on regrowing origin as they are gained from rapeseed oil. The polarity of OCT is suspected to be quite low because of its monovalent alcohol.

- *polyalphaolefin* (PAO) was chosen as a low polarity reference oil candidate. It is a completely mineral based group IV oil.

The viscosities of the base oils were selected within a narrow band to prevent a high deviation of the thickener base oil ratio within the grease matrix. The main physicochemical base oil characteristics are collected in table 4.1.

Table 4.1.: Base Oil Characteristics

baseoil	$\nu_{40}[\text{mm}^2\,\text{s}^{-1}]$	$\nu_{100}[\text{mm}^2\,\text{s}^{-1}]$	density$[\text{g cm}^{-3}]$	degradability[%]
HOSO	38,8	8,5	0,92	96
TMPO	48,0	9,8	0,92	96
OCT	25,5	5,5	0,87	96
PAO	46,7	7,9	0,83	yes

4.1.1.2. The thickeners

For the formulation of candidate and reference greases, thickener types that generally qualify for the production of bio greases have to be selected. As was made clear in the state of scientific research, the field of biogenic thickeners is yet to be investigated more thoroughly to bring forth commercially available biogenic thickeners. For this reason traditional thickeners that, although not biogenic, qualified for the production of bio-greases were chosen. These are *highly dispersed silica acid* (HDS) and *Bentonite* (BT) as representatives of the group of inorganic mineral clay thickeners as well as two soap thickeners, a *calcium soap* (Ca) based on Ca-12-hydroxy stearate and a *lithium soap* (Li) based on Li-12-hydroxy stearate, which both prove to be partially biodegradable [7]. Although Li-soap would not be the preferred thickener for bio-greases it was still chosen for reasons of referencing with traditional thickener types. As previously mentioned all greases were targeted to an NLGI consistency grade 2. This resulted in different thickener concentrations due to slightly different base oil viscosities. Compositions and physical grease characteristics are depicted in table 4.2.

Although compliance verification has not been furnished officially, the formulated bio-greases investigated in this study theoretically fully comply with the above-mentioned criteria of the Blue Angel. This eco label was used only exemplarily for many other labels with similar requirements; other comparable eco labels can be found elsewhere [102] and are discussed along with recommendation to their test methods [103, 104].

4.2. Selection of experimental procedures

4.2.1. Polarity measurements and equipment

In previous studies [51, 105, 106], oil polarity is determined by means of measuring the polar shares of surface energy of the oils with the application of equations by either Wu [107], Owens, Wendt [65] or Fowkes [108–110]. The current state of research suggests several ways to obtain the polarity of oils. Some authors make use of solubilization methods [111], others use gas- and other chromatographic methods like i.e. HPLC [112–115] and interfacial tension measurements between oil and water [116] in order to obtain oil polarities. El-Mahrab-Robert et al. [117] compare some of these methods with a result of superiority of the two latter-mentioned ones. In this study the relative polarity of the given base oils was determined by interfacial tension measurements between oil and water using equivalent equipment and methods to the ones described in [116] and [117], using a K100 Ring tensiometer designed by *Krüss GmbH Wissenschaftliche Laborgeräte, Hamburg, Germany*. For the preparation of the measurements, a suitable amount of distilled water has to be filled into a measuring glass, and after complete emersion of the prior cleansed PtIr ring a sufficient amount of oil is added to the water forming an interfacial surface. The surface tension of the same was measured a suitable amount of times repeatedly, after which the arithmetic average of all measurements was calculated.

Oil polarities measured in the described manner were not determined as a specific quantity with a corresponding physical unit. Rather, the results of such kinds of interfacial tension measurements can only bring to light a relative comparison of the measured oil polarities. The general way to interpret these results is to attribute high interfacial tensions to low polarities. In other word, the lower the interfacial

4. Materials, methods and conditions for the experimental analysis of greases

Table 4.2.: Grease Composition and Some Physical Grease Characteristics

Grease label	Thickener Type	Thickener Content [%]	Pu [1/10 mm]	Pw [1/10 mm]	NLGI Class	DP [°C]	oil separation [wt%]
HDS PAO	Highly dispersed silica	10,3	269	280	2	-	1,12
HDS HOSO		13,9	238	268	2	-	0,92
HDS OCT		11,4	264	287	2	-	1,22
HDS TMPO	acid	14,1	249	287	2	-	0,82
BT PAO	Bentonite	20,1	249	279	2	-	-
BT HOSO		28,5	257	268	2	-	-
BT OCT		15,8	264	291	2	-	-
BT TMPO		24,0	242	279	2	-	-
Ca PAO	Ca-stearate	18,8	283	287	2	133	1,16
Ca HOSO		17,6	272	279	2	132	0,80
Ca OCT		15,7	253	272	2	136	0,88
Ca TMPO		13,7	260	272	2	132	0,63
Li PAO	Li-12-Hydroxy stearate	9,7	279	283	2	205	2,57
Li HOSO		15,9	264	272	2	195	1,12
Li OCT		13,6	279	294	2	188	1,79
Li TMPO		20,7	264	279	2	190	0,81

4.2. Selection of experimental procedures

tension between a measured oil and water, the higher is its polarity.

4.2.2. AFM imaging and equipment

The greases whose soap structures are depicted in figure 2.2 were prepared according to either of the methods described by Kistler [118] or Farrington [38] prior to the SEM imaging process. A big disadvantage of these methods and those described in [119–122] is the risk of harming or physically changing the soap fibers in the process of freezing or washing out the oils. Another disadvantage is the fact that soap fibers in the SEM or TEM images taken in the described manner are not displayed in their actual situation surrounded and completely encompassed by oil. For these reasons SEM images may be deceptive images of reality. With modern micro imaging methods it is possible to investigate the soap structure without prior extraction. Sanchez et al. [123] used the method of atomic force microscopy (AFM), which offers a substantial advantage compared to other micro-scanning imaging methods, such as SEM or TEM. This method, is advantageous because the thickeners of the investigated greases do not need to be isolated, for instance by oil removal with a chemical solvent nor do the samples need to be frozen. Hence, this method is also applied in the studies performed in this work.

Prior to the imaging process the greases were heated to temperature conditions close to the dropping point of the respective grease for just a short period of time. Subsequently, they were cooled down forming an even surface, which can be scanned on the micro scale. Thus prepared, the samples were placed under the scanning probe of the AFM (*Digital Instruments, Veeco Metrology Group Inc., Santa Barbara, CA*), which subsequently scans the grease surface mechanically in the tapping mode. In this scanning mode the AFM tip oscillates at a frequency close to resonance with relatively high amplitudes as compared with other scanning modes. While scanning, force interactions between the surface to be investigated attract the scanning probe and thus change its oscillation amplitude and result in a phase shift between excitation and natural frequency of the probe cantilever. An actuator adjusts the distance between the probe and the scanned surface and thereby the force interaction. All this data is interpreted into an image of the scanned surface.

4.2.3. Tribological tests and equipment

Some tribological investigations are necessary to prove the tribological assumptions made in 3.1. As the assumptions pertain to fundamental tribological mechanisms they will be elucidated by fundamental friction and wear tests much better than by highly sophisticated specific tests based on market specifications and applications. Therefore, only fundamental friction tests will be performed. More important than the size of the portfolio of tests to be performed is the choice of the right tests suitable to find the presumed effects. In this background, a small number of suitable tests will be sufficient if the type and intensity of friction and wear mechanisms encountered in the tests lend themselves to disclose the desired effects.

Fundamental tribostests are far away from market applications and were introduced to simulate frictional and wear behavior. In order to achieve quick results only tests with sliding friction in the state of mixed friction were performed. The load, relative

4. Materials, methods and conditions for the experimental analysis of greases

velocity of contacting surfaces and their contact geometry must be adjusted to the state of mixed friction.

The nanotribometer by *CSM Instruments SA, Peseux, Switzerland*, shown in section 4.1, was used for tribological friction and wear tests of the selected greases. This ball-on-disc apparatus can work in linear oscillation and circular rotation mode.

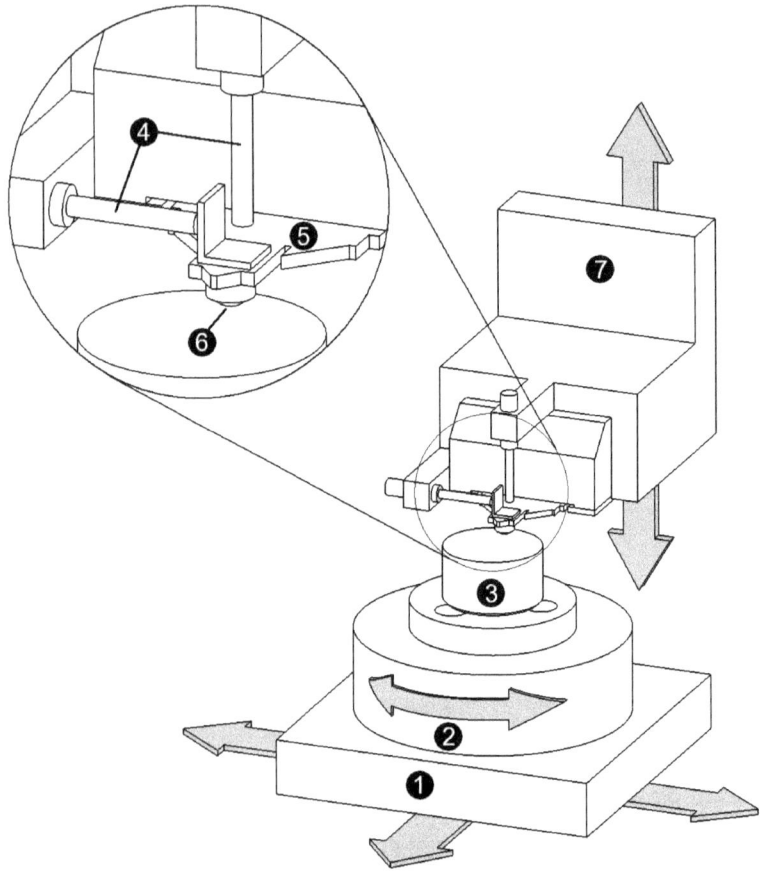

Figure 4.1.: Experimental setup of the nanotribometer—Note:1 XY-cross-table, 2 Rotational axis, 3 Specimen steel disk, 4 Interferometrical sensors, 5 Cantilever, 6 Specimen sapphire ball, 7 Z-step-motor-axis with integrated piezo-axis

In this work, all tests were performed in the rotational mode. The velocity of relative motion between the static ball and the dynamic disc directly depends on the diameter of the set wear track in correlation to the set rotational speed resulting in sliding friction. The step motor of the rotational axis makes up to 120 rpm and is mounted to an XY-cross-table, which positions the disc under the ball in order to set up a controlled track radius. Specimen balls were fixed to a cantilever module and normally positioned to the plate by a Z-step-motor-axis. Defined load was applied to the ball by fine positioning and deflecting the cantilever via piezo-axis. The deflection of the cantilever is interferometrically measured and then controlled. By means of two distally orthogonally attached mirrors the deflection of the cantilever in the radial (frictional force) and normal (normal force) directions of the plate revolution is detected and, by very precise calibration of the spring constant of the cantilever in both main directions, calculated into frictional and normal forces. Both of these forces, the friction coefficient, the relative indentation of the ball into the plate as well as the ambient temperature and the relative humidity of the measuring cabin are detected and logged during the measurements at a set sample rate. The nanotribometer only works at ambient pressure and temperature without temperature or pressure control. The utilization of cantilevers with different spring constants covers a large load range from 0.1 up to 500 mN. The high-precision optical interferometric measurement detects the wear height with a resolution of 50 nm. Despite of relatively low normal forces of the nanotribometer high values of Hertzian stress (up to $\sim$2.0 GPa) can be applied to the measured tribological system. This is due to the fact that very small balls may be applied to the cantilever.

4.2.3.1. Tribological test parameters and materials

As discussed earlier, the tribotests selected to investigate the assumptions are all performed in a sliding contact situation of a spherical and a plane surface. All tests were performed with balls of 1.5 mm diameter and in rotational mode with a relative sliding speed of $5.0\,\mathrm{mm\,s^{-1}}$, which in this combination results in mixed friction contact. The normal load is varied with values of 500 mN, 100 mN, 10 mN and 1 mN to make sure that the state of mixed friction is investigated. This wide range of normal loads in combination with different materials results in different Hertzian stresses depending on the material combinations.

In order to prove the previous assumptions it is necessary to not only select base oils of different polarities but also tribological contact partners of different materials. Therefore the material combination is varied in the tests, using either a polycrystalline sapphire and steel balls 100Cr6 that is pressed against steel plates (115CrV3, hardness 22.72 HRC). The mentioned sapphire was selected because of its covalent bonding character with low tendency towards the formation of dipoles in the surface structure. The mentioned steel plates and balls were selected because of their well known application in tribological applications. All steel plates were metallographically grinded and polished with a soft finish diamond paste of particle size 3 µm in the last polishing step to keep the tribological results on a comparable level. All sapphire surfaces were used only once for tribological characterization before replacement.

The tribological test evaluation was performed with the controller software of the nano tribometer. All data logged was taken into consideration. The wear marks in the balls and the discs were investigated with different optical metallographical imaging processes.

4.2.4. Rheological tests and equipment

Only standard rheological tests need to be performed for the investigation of the selected greases. As shown in the state of current research these are very well qualified to show the searched effects and to prove the assumptions made in section 3.1. Selected test were performed to investigate the rheological postulations. These are rotational tests to elucidate the transient shear flow of all greases, oscillatory amplitude sweep tests and frequency sweep tests—all of which as described in the current state of research. Because of the temperature dependence of some rheological effects postulated, a complete series of rheological investigations needs to be performed with varying temperature conditions. Transient shear flow tests were selected for this purpose. All rheological tests of the above-mentioned greases will be carried out on the modular compact, controlled-stress rheometer *MCR300, Anton Paar GmbH, Graz, Austria*.

4.2.4.1. Rotational transient tests

The rheological rotational transient investigations of the aforementioned greases (see section 4.1.1) were carried out on the modular compact, controlled-stress rheometer MCR300 described in section 4.2.4. Rotational tests were performed using a plate-plate system with a diameter of 25 mm, applying a constant shear rate of $\dot{\gamma} = 1 \mathrm{s}^{-1}$. All samples were given enough time to relax from the shearing stress during the setting the gap and to adjust to temperature settings before the measurement was started. The measuring time was set to 2000 s, which at $\dot{\gamma} = 1 \mathrm{s}^{-1}$ is enough time to reach the aforementioned limiting steady-state shear stress. All tests were performed at different temperatures (-10 °C, 25 °C, 40 °C and 80 °C). Every measurement of the 16 greases was repeated three times for each temperature.

4.2.4.2. Amplitude sweep tests

All greases described in section 4.1.1 were examined in the oscillatory shear mode with inclining amplitude (amplitude sweep) using the MCR-300 rheometer described in section 4.2.4. All measurements were performed at a temperature of 25 °C with a plate/plate system and a gap of 1 mm as recommended in DIN 51810-2. The measurement was parametrized to an inclining deformation starting at 0.01 % and ending at 100 % with a logarithmic ramp. The frequency of the strain oscillation was set to 1 Hz. Each measurement was provided with 20 min of relaxing time after setting the gap. This time period was judged to be enough time to also adjust to the preset temperature. Each measurement of the sixteen greases was performed four times and averaged before evaluation in order to reach a statistically safe reproducibility.

4.2.4.3. Frequency sweep tests

Since the shear stress τ [Pa] was selected as the actuating variable for the constant amplitudes in frequency sweeps, the corresponding stresses had to be determined for the selected save range of deformations inside the LVE-range. This was done

4.2. Selection of experimental procedures

prior to the test using small amplitude oscillatory shear with amplitude sweep. The values of applied shear stress are displayed in table B.1 in the appendix B.

All measurements were performed in a plate/plate system with a gap of 1 mm. The rheometer was parametrized to wait for 20 min after setting the gap in order to allow the grease to relax after the stress applied during attaining the measuring gap and to adjust to the preset temperature of 25 °C. A logarithmic frequency ramp starting at $100\,\text{rad s}^{-1}$ and going down to $0.03\,\text{rad s}^{-1}$ was performed. A three-fold repetition of the measurements secured reproducibility.

4.2.4.4. Tensile tests

Tensile tests as described in section 2.5.4 were performed with all greases examined in this work. All of these tests were performed with the MCR 300 rheometer illustrated in section 4.2.4. All tests were performed at 25 °C with varying presets of the gap in 1 and 0.5 mm hight. For reasons of reproducibility all tests were performed three times. In the preparation of the tests a suitable amount of grease was placed in the measurement system of the rheometer, the gap was set to the preset value and excess amounts of grease were scraped off. The velocity for increasing the gap was set to 0.25 mm /s for all tests and the resulting normal force was monitored collecting 50 measuring points per second within the first two seconds after the start of the opening. After this starting phase the data collection rate was set down to only 10 measuring points per second to limit the amount of data generated from each single measurement set.

4.2.5. Other test equipment

The only tests that could not be performed on standard equipment are bouncing ball tests as described in section 2.6.1. For this reason a special rig was designed with adjustable hight of fall and an adjustable high speed camera socket. The surface samples were fixed to granite plate and loaded with metal collar of 1.0 kg of weight to prevent them from bouncing off the granite plate after the impact of the steel balls.

4.2.5.1. Bouncing ball tests

Bouncing ball tests, as introduced in section 2.6.1, were performed with all given greases. Standard roller bearing balls 100Cr6 with a diameter of 12.7 mm and a weight of 8.42 g were dropped from a preset relative hight of 0.3 m over bottom level. For reasons of referencing, the same metal samples (115CrV3) as applied in previously reported tribological investigations were used as the impacted surfaces. All samples were evenly polished with a diamond soft finish of 3 µm just as was done in the tribological tests in the nano tribometer. Grease layers of 0.3 mm were applied to each surface prior to the tests. The metal samples were clamped into a jig at bottom level before the balls were dropped. The length of the grease filament created underneath the ball during bounce off were measured with a high-speed camera with a sample rate of $1200\,\text{frames s}^{-1}$. Each test was repeated six times and all imprints in the polished metal surfaces were microscopically examined. After the tests the surfaces of the samples were ground and polished for reuse.

5. Experimental analysis

5.1. Polarity analysis

Polarity measurements of the base oils were conducted with the equipment and methods described in section 4.2.1. The interfacial tensions between all investigated base oils and distilled water were measured ten times repeatedly within a time period of 20 min in the aforementioned manner. Averaged results of these measurements are displayed in table 5.1.

Table 5.1.: Interfacial tensions measured between all investigated base oils and distilled water

Baseoil	PAO	HOSO	OCT	TMPO
Interfacial tension [mN/m]	20,0	9,0	23,0	7,5

According to the valid interpretation of interfacial tension measurements, these values result in a relative ranking of polarities of the given base oils in the following order: OCT≈PAO≪HOSO≈TMPO. These results are substantiated by the molecular structure of the investigated oleochemicals. Both, HOSO and TMPO are constituted of mainly triglyceride esters known to exhibit high electronegativities, which evoke a higher propensity to create permanent dipoles. Biologically rapidly degradable oils, especially esters of glycerine with saturated and unsaturated long chain fatty acids are known for their very high relative polarities [51, 64, 106]. The O-C bonds of esters and the O-H bonds of some fatty acids and alcohols in oleochemicals evoke a large charge imbalance caused by different electro negativities. This results in the formation of permanent dipoles [124], high tendencies towards the formation of Keesom-forces and consequently high polarities of oleochemicals. This, however is not the case for PAO and especially OCT.

The polarities of the tribologically investigated surfaces in this study are taken from relevant literature [125] and can be ranked as $Al_2O_3 < 115CrV3 \approx 100Cr6$.

5.2. AFM analysis of the greases

All selected greases underwent microscopic examination by method of AFM analysis as described in section 4.2.2. With all the mentioned advantages of AFM investigations, there are also unfavorable circumstances of this method, which emerged especially in the examination of Ca greases. The process of heating up the greases to a temperature region just beneath their dropping points, in order that their surface be smoothened enough for the AFM analysis, led to thermal degradation of the Ca thickeners. For this reason it was not possible to receive any useable AFM image

5. Experimental analysis

of Ca-thickened greases. With all of the other thickening agents, however it was possible to gain AFM images of more or less good quality. A selection of these images is displayed in figures 5.1 to 5.5 on the following pages. The thickeners of

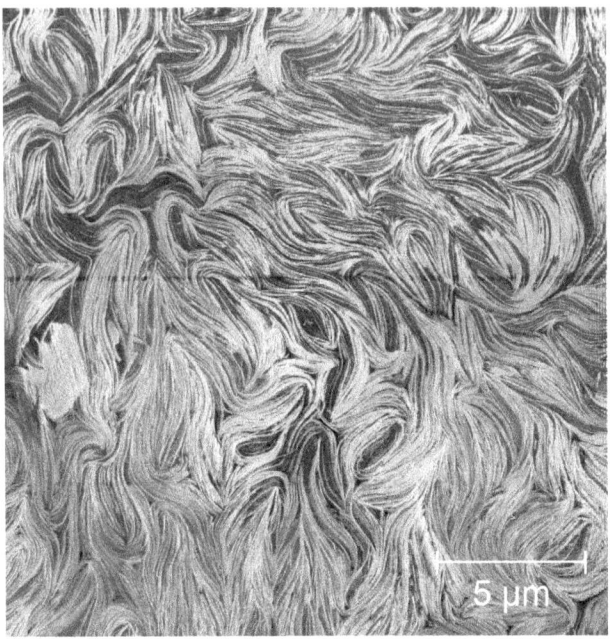

Figure 5.1.: AFM image of Li-HOSO grease

the greases appear as bright to white areas in all displayed AFM micrographs. The brighter an area appears in the AFM image the more rigid is the scanned part of the specimen. Dark areas in the AFM image analogously denote soft or even liquid parts representing the enclosed base oil. It should be mentioned that the rigidities of the displayed AFM images are not to be compared to each other because each image was manually normalized to picture the full spectrum of gray scales. Still, the structure and the composition of the grease thickeners appear very clearly.

The AFM image depicted in figure 5.1 was taken with a Li grease in combination with HOSO base oil. This image very clearly depicts the fibrillar structure of the Li thickener with single, partially interwoven and entangled loops of different size. As the AFM only scans a superficial layer of the grease specimen the lengths of the single fibers can only be estimated very vaguely. But still this image clearly shows the partially homogeneous and parallel arrangement of the soap fibers. This figure very well shows the arrangement of the soap fibers within a fully formulated grease as opposed to those micrographs taken by method of SEM, as depicted in figure 2.2. In a direct comparison one notices the clear advantages of the AFM micro imaging methodology as previously mentioned. The mutually interacting mechanisms of attraction between oil and thickener structure result in this typical fiber arrangement. It is presumed that the attraction between the soap fibers that partially aligns them parallel to each other is a direct consequence of the polarity of base oil molecules.

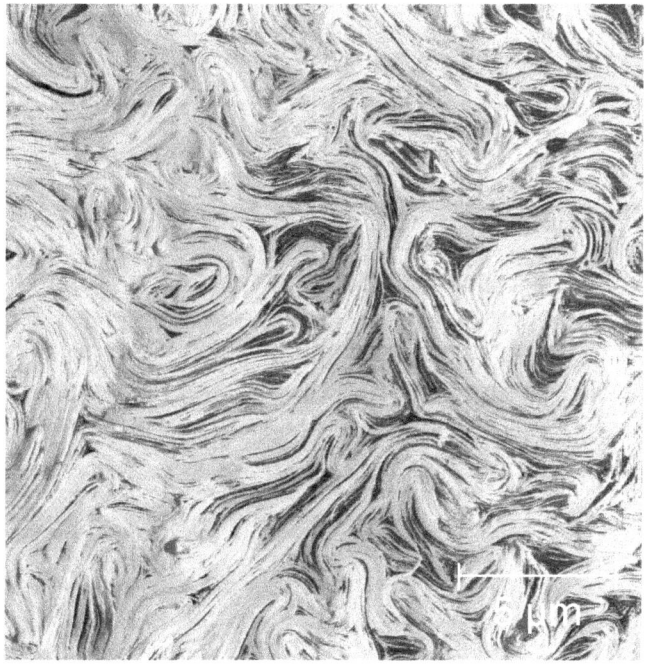

Figure 5.2.: AFM image of Li-TMPO grease

Figure 5.2 depicts the AFM image of a Li grease with TMPO base oil in the same scaling as the previous image. This figure also shows the interwoven soap fiber loops encompassed by the base oil. The length of the soap fibers appears slightly longer than in the previous figure, whilst all other attributes such as the homogeneity of the partialy parallel arranged fibers seem alike. The interwoven fiber bundles appear a bit wider than in the micrograph taken from a Li-HOSO grease.

The arrangement of soap fibers in figure 5.3 displaying an AFM image of a lithium grease based on OCT oil appears much different. Here the loops of the soap fibers, although arranged in a similar manner, are much smaller in diameter. In this image too, the lengths of the soap fibers can only be estimated vaguely as it is hard to detect where one fiber ends and another fiber begins. Also, the fibers appear much narrower in combination with Li-OCT as compared to Li-TMPO and Li-HOSO. The most obvious difference between the soap fibers of Li-OCT in figure 5.3 and those of Li-HOSO and Li-TMPO in previous figures 5.2 and 5.1 pertains to the rate of entanglement of the soap fibers—meaning the number of loops per examined unit of grease volume.

5. Experimental analysis

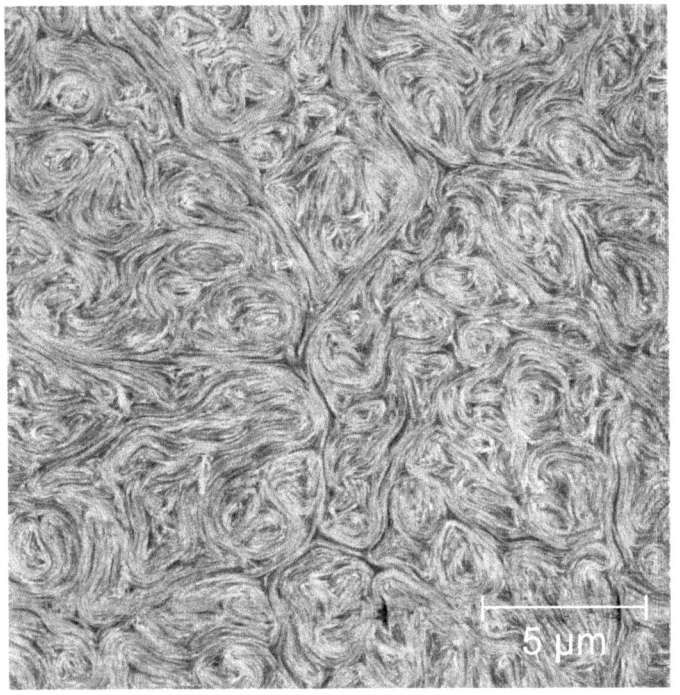

Figure 5.3.: AFM image of Li-OCT grease

5.2. AFM analysis of the greases

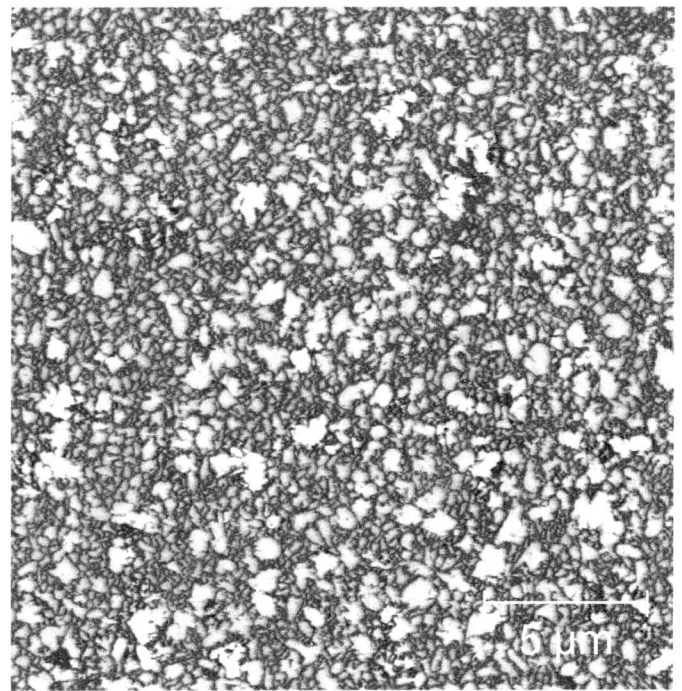

Figure 5.4.: AFM image of BT-HOSO grease

5. Experimental analysis

The AFM image of BT-HOSO grease displayed in figure 5.4 very well reveals the nature of BT thickeners in a fully formulated grease. The montmorillonite particles described in section 2.1.3 appear as evenly distributed solid platelets easily spotted in white to bright gray scale areas in figure 5.4. These particles are surrounded by the liquid base oil, which appears as dark gray to black regions. One not only notices the size of these platelets but also how densely and multi directionally they are arranged—some are arranged parallel to the scanned superficial layer, some normal to it. Figure 5.4 also makes clear that the size of these clay particles in BT thickeners varies starting from small fragments not larger than 0.1 µm up to platelets of around 1 µm in diameter. In the process of normalizing the image to cover the full spectrum of grey scales it was noticed that the difference of hardness between oil and platelets is much bigger than the hardness difference between oil and Li soap fibers in the previous AFM images. Although this difference can not be expressed in numbers it still indicates the extreme hardness of BT particles.

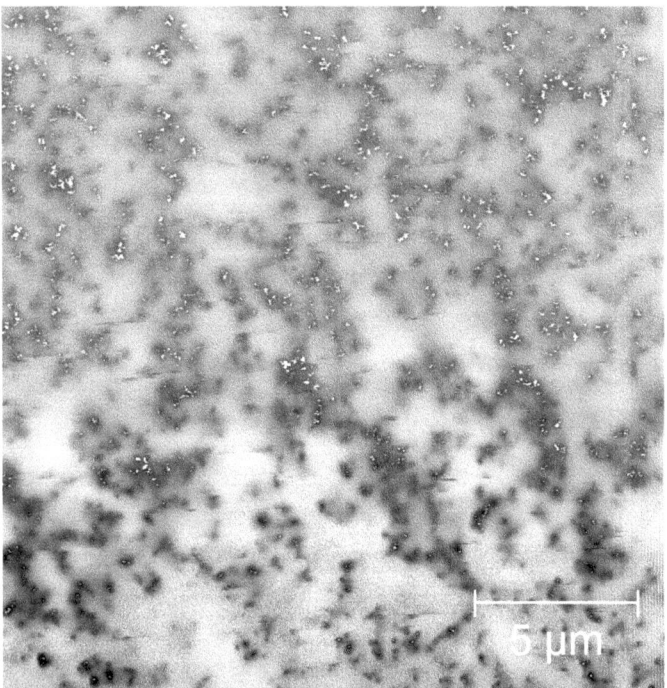

Figure 5.5.: AFM image of HDS-OCT grease

Figure 5.5, on the other hand, delineates the completely different microstructure of an HDS-OCT grease. This figure clearly shows relatively large grayish areas, that indicate to an evenly distributed solidity of the examined substance according to the rules of evaluation of AFM images. For one thing, a homogeneously distributed thickener structure can be derived from this evenly distributed solidity. For another thing, this points to the very small seize of HDS particles that mostly consist of SiO_2 [41]. With this detected homogenous structure it is not surprising that all examined HDS greases are translucent substances. Additionally, evenly distributed white spots of high hardness conspicuously appear in figure 5.5. These are surrounded by very dark edges that indicate to fluid characteristics. The brightness of these spots is evidence of a much harder structure than of the just-described grayish areas of homogeneously spread thickener structure. It is presumed that these bright spots represent agglutinated thickener particles. Based on the fact that all these spots are surrounded by a fluid characteristic edge it is presumed that agglutinated thickener particles lose their ability to physically tie the base oil.

The depiction of further AFM micrographs of BT- and HDS-based greases is omitted for lack of expressiveness. With other images taken it was found that the base oils only insignificantly influence the microstructure of these clay thickeners.

The fact that polarity influences on the rheological behavior of greases have been postulated suggests the assumption that they may also influence the formation of the thickener structure. If at all, this effect should only be found in combination with soap thickeners as clay thickener particles are already chemically build up in advance to the grease formulation and therefore they remain chemically unchanged. The found difference of the rate of entanglement is much higher in the Li grease based on OCT oil than in those based on TMPO and HOSO. It is presumed that this fact and all other differences are based on the base oil polarity as postulated in section 3. Longer, wider and less entangled fibers with less loops per examined unit of volume result from the combination with highly polar oils TMPO and HOSO as compared to the low polarity base oils OCT and PAO. In addition, it is found that this effect also exerts an influence on the thickener/base oil ratio in combination with Li thickeners as shown in table 4.2. These found results confirm the formation of model B as postulated in section 3. Further investigations will show if the effect also influences the rheological behavior depending on base oil polarity.

The proven fact that the soap fiber structure is a function of the applied metal soap, the type of microscopic visualization and the base oil polarity suggests that it may also be influenced by other processes. This presumption is confirmed by other authors who found a dependence of the pre-treatment [120, 126] and processing conditions [28, 127] on microstructure and rheological behavior. Also, previous studies [21, 22, 128] found a dependence of soap structure on the rheological shear history.

5.3. Tribological analysis

All of the aforementioned greases (see section 4.1.1) were tribologically examined with the nanotribometer described in section 4.2.3. In the test procedures the normal force was varied with resulting Hertzian stresses depending on the material combinations. A grease layer of 0.05 mm was applied to the steel plates. Then, in the second step the measurement parameters of normal force, relative speed and track radius were set up. The test duration was set to 50 min each without manual re-lubrication of the steel plates. Because this fact may involve a danger of critical grease deficiency in the contact the development of friction coefficient over time was

5. Experimental analysis

observed thoroughly to ensure that no starvation effects occurred. In this way, it is guaranteed that tribocontacts were always fully flooded and operated under steady state condition. In order to allow conclusions to be drawn about the friction and wear influence of each bulk component of the greases the above-described tests were performed with all greases, each base oil on its own and in completely dry contact situation. The results will be discussed separately in the two subsequent sections divided into groups of different material combinations.

5.3.1. Metal soap-, HDS- and BT-greases in sapphire-steel contact

The preset normal forces of the tests resulted in values of Hertzian stress depending on the material combination. The given combination of sapphire balls and steel plates resulted in values as shown in table 5.2.

Table 5.2.: Normal forces and resulting values of Hertzian stress in the nanotribometer—sapphire balls and steel discs

normal force [mN]	500	100	10	1
max. Hertzian stress [GPa]	1.94	1.13	0.53	0.24

5.3.1.1. Friction results

Generally, friction coefficients vary with high fluctuation in most frictional tests. Averaged coefficients of friction of all examined greases in the selected material combination sapphire ball on steel plate are shown in figure 5.6 in logarithmic scale. This bar diagram clearly shows that the friction coefficient is influenced bynormal

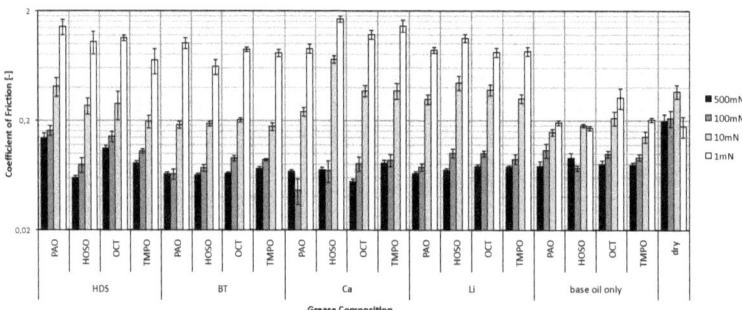

Figure 5.6.: Friction coefficient in contact situation sapphire ball on steel disc—all greases studied, pure base oils and dry contact situation

load with a general tendency towards much higher values for low normal loads. In fact, values increase up to tenfold when comparing the results of 1 mN to those achieved with 500 mN. An explanation for this effect might be found in the set-up of the experiment. All greases were applied to the steel discs only prior to the start of the test, resulting in the fact that the static test ball had to plough its way through the grease layer in every revolution of the plate. This ploughing resulted in grease layer displacement, evoking a rheological resistance to the relative motion, which, although steady for all grease layers of comparable consistency and thickness, is insignificantly small at high normal loads but the main resisting factor for very small normal forces. In other words, thicker grease layers evidently cause higher resisting abilities towards repositioning by the specimen ball. For this reason, it was very important to always apply grease layers of the same thickness to the steel discs. In the completely dry contact situation and in the tribosystems lubricated with pure base oils this difference is not as severe as in the grease lubricated systems—a fact, which substantiates this theory of rheological resistance.

Frictional results in the present work have an average standard deviation of more than 10%. Besides this, fluctuations in values of coefficients of friction appear a bit lower for higher normal forces of 500 and 100 mN than in the lower normal load region of 10 mN and less.

Since the pictured behavior of the development of friction coefficients in low normal load situations is too much influenced by rheological grease characteristics the focus of evaluation will be drawn to the higher normal loads 500 and 100 mN. In attempting to attribute frictional values to bulk grease components one has to regard results from each component's point of view. At first consideration from the thickener's point of view one detects a general tendency of highest frictional values for the series of HDS-thickened greases with an average of 0.110 ± 0.036. Especially HDS-PAO reaches friction values that are comparable to those achieved in the dry contact. The group of calcium-thickened greases stands out due to its lowest friction coefficients in comparison (0.070 ± 0.014), followed by BT (0.073 ± 0.011) and lithium-thickened greases (0.081 ± 0.014).

When comparing component attributes and their influence on friction from the base oils' point of view one first of all finds out, that the ranking within the group of base oils from PAO with highest frictional values down to HOSO with the lowest coefficients of friction is also found in the group of greases formulated with HDS-thickener with this tendency even more intensified. Reasons for this effect will be discussed later on with respect to wear intensities and mechanisms. It also shows that the group of pure base oils presents slightly higher frictional values than Ca-soap thickened formulations in the present frictional contact situation, sapphire ball on steel plate. Moreover, the ranking that was found for base oils is almost completely turned to the opposite tendency now with least coefficients of friction for Ca-PAO. Results also show that influences of single base oils on Li- and BT-thickened greases are too small to be detected or even interpreted into tendencies.

5.3.1.2. Wear results

Different behavior of grease formulations and their components are generally more explicitly displayed in wear responses than in frictional results. In the same way, wear measurements fluctuate less than frictional measurements. Considering the full

5. Experimental analysis

dimensions of wear only after the completion of a frictional measurement implies to break a process factor down into a static view. This way fluctuation is eliminated to some extent [129]. But it is also important to keep in mind that friction and wear always depend each other in the total frictional system. At the examination of friction and wear as two interrelating dimensions it is insufficient only to take frictional values and wear dimensions into consideration in order to receive significant information about the tribological system. When trying to understand and correctly evaluate friction and wear one must also comprehend their states, types and especially their mechanisms. Different wear states, evoked by specific wear types are made manifest through different wear mechanisms, which take a deep impact on frictional and wear behavior [74, 75]. As a consequence, different wear mechanisms result in different wear dimensions but also in altering frictional values. Moreover, it is possible to even have several different wear mechanisms present in one tribological contact system. Micrographic appraisal of wear results in the retrospect of tribological tests gives direct indication on wear behavior, which consists of the aforementioned wear mechanisms and wear types.

In this series of experiments wear was analyzed statically and process orientated. For the static part, which was measured after the tests, the widths of all wear tracks on steel discs and the corresponding diameters of all wear scars on sapphire balls are displayed in a diagram for all measured normal forces in figure 5.7. For a better

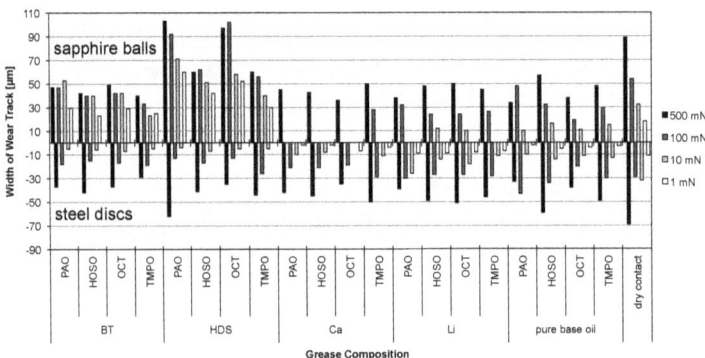

Figure 5.7.: Widths of wear tracks in sapphire balls and steel discs—all greases, pure base oils and dry contact

overview and comparability the values of the wear tracks in the discs were negated and their bars colored in the same grayscale as those of the corresponding wear marks in the balls. Wear responses will be discussed along with the information provided by micrographs of some selected tribosystems assessed in the course of the experiments. These representative wear marks are displayed in figures 5.8 to 5.12. Each individual figure exhibits the micrographs of wear results of a specific tribosystem in both the disc and the ball examined at all given normal load conditions (1 mN up to 500 mN). As it has been done previously with the interpretation of frictional results,

5.3. Tribological analysis

wear characteristics are also associated with all single bulk grease components. The images show that ball and disc wear need to be evaluated separately. Some greases e.g. produce high wear rates in the ball but leave the discs unaffected and vice versa. It also becomes apparent that each normal force has its own characteristic influence on wear behavior so it does not always prove to be true when simply predicting less wear as a consequence of less normal force. As the diagram in figure 5.7 and the aforementioned micrographs of selected tribosystems show that bulk grease components highly influence the wear response on the tribological systems, wear results are discussed in the following paragraphs by referencing to the same. Since thickener influences on wear responses are more clearly exhibited and different base oils merely intensify or reduce this influence, rather than changing its mechanisms, the discussion will primarily be sorted into groups of thickeners.

HDS-greases in sapphire/steel contact The most evident fact for HDS-based greases is that they produce the widest wear scars in sapphire balls in all of the investigated greases as exhibited in figure 5.7. In some instances they are even wider than in completely dry contact situation. At a normal load of 500 mN this same effect takes place on the surface of the steel plates, resulting in wear tracks just as wide as those measured with pure base oils with values ranging from 35 µm in case of HDS-OCT up to 65 µm for HDS-PAO. Only the non-lubricated dry contact resulted in slightly wider wear tracks in the discs (69 µm). The smaller the normal loads, however, the more diverge these tendencies. While wear scars on the sapphire balls remain the biggest compared to all other lubricants measured, and even compared to dry contact, the wear tracks on the steel plates become smaller and smaller. This effect becomes most evident for 1 mN, the smallest of all normal forces applied in this series of experiments. At this load HDS-thickened greases still produce very wide wear scars in sapphire balls ranging from 30 µm in case of HDS-TMPO to 52 µm in diameter for HDS-OCT (displayed in white solid bars in the diagram in figure 5.7), while on the other hand they do not produce any detectable wear on the steel plates. This effect also occurred in the group of BT-greases. Another detectable effect also known from the group of BT-greases is the clear division into two almost equal subcategories (PAO/OCT and HOSO/TMPO). This division, in contrast to BT-greasess, is much more distinct in the group of HDS-greases with PAO/OCT reaching an average wear width of 79.4 µm and HOSO/TMPO resulting in 50.1 µm. In the group of HDS-greases this effect is also consistent throughout all applied normal loads.

The group of HDS-greases tested under all normal load conditions and represented by an exemplary tribocontact lubricated with HDS-OCT grease in figure 5.8 presents itself with very peculiar wear marks. The wear marks produced on the discs in the given sapphire/steel contact situation also show a highly abrasive character in the central contacting area. Wear tracks also disclose a high proportion of plastic deformation when applying high normal forces (500 and 100 mN), resulting in a deep groove with high flanks at its borders. The application of 1 mN normal load results in no detectable wear at all. The sapphire balls as counter parts in this contact also show a high rate of abrasive material removal from the surface, visible as a circular worn off cap. Flanks, which showed as elevations in the discs, appear

5. Experimental analysis

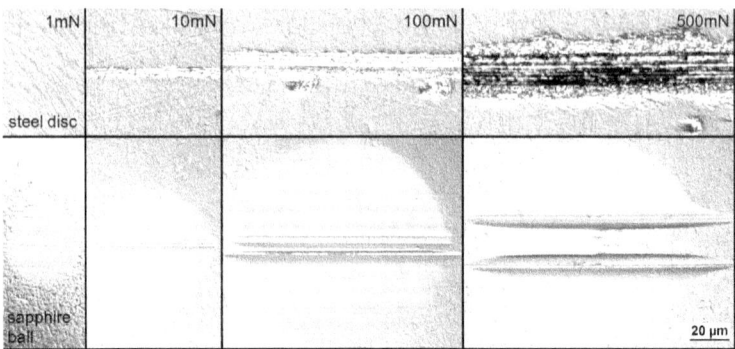

Figure 5.8.: HDS-OCT in material combination sapphire ball on steel disc

as corresponding grooves in the sapphire balls. More interesting, however, is the fact that the rest of the circular worn off area appears as a smooth, almost evenly polished region, although the corresponding counter part of the steel disc contact appears completely unspoiled. This circumstance seems even more abnormal when considering the hardness of sapphire compared to the hardness of the steel discs. Ensuing analyses of all lubricating greases are recommended to elucidate mechanisms that favor this effect. In order to further investigate this very peculiar kind of wear mechanism this test was repeated with equal conditions in normal load of 500 mN and HDS-OCT as lubricant. The resulting wear marks in the sapphire ball were optically scanned with white light interferometry as depicted in figure A.1. In addition to a perspective spacial view this figure also shows two surface scanning profiles in the directions parallel to the relative motion between the ball and the disc orthogonal to the motion. The cross-section view normal to the direction of motion exhibits a sharp edge of 1 μm hight with two proximate grooves, the deepest of which reaching 0.5 μm beneath the level of the main worn off cap. The rest of this curve and the cross-section parallel with the direction of motion reveal the smoothness of the surface of the worn off cap with asperities of hight in sub-micrometer level.

BT-greases in sapphire/steel contact The most evident thing in the group of BT-greases are the steady wear widths in the sapphire balls in almost all applied normal loads as found in figure 5.7. All wear widths of BT-PAO, -HOSO and -OCT in the normal load range between 10 and 500 mN lie within a narrow spectrum between 40 and 53 μm. Only BT-TMPO resulted in slightly narrower wear widths with less normal load applied. This figure also depicts that, in the normal load situation of 500 mN, the widths of the wear tracks in the discs correspond to those in the sapphire balls. This is only true for 500 mN, however—in smaller normal load conditions the wear tracks in the discs are much narrower than the corresponding marks in the balls. In the normal load condition of 1 mN this effect even leads to wear marks in the balls, with a range from 23 μm in case of BT-HOSO up to 30 μm in case of BT-PAO, that are even wider than the ones produced in non-lubricated

5.3. Tribological analysis

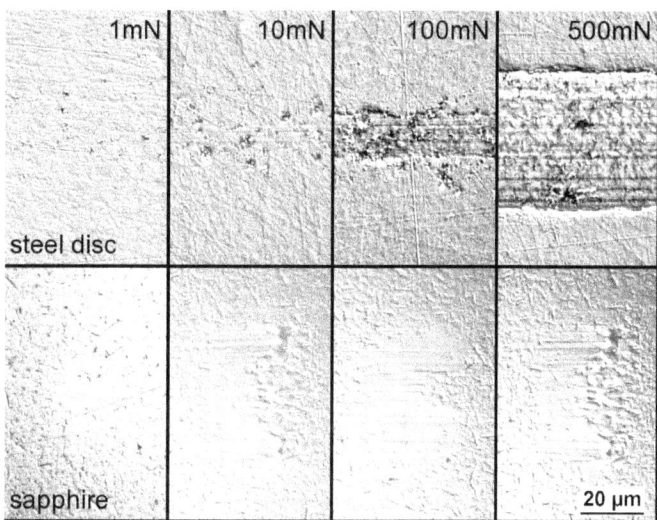

Figure 5.9.: BT-HOSO in material combination sapphire ball on steel disc

dry contact (18 µm) while the steel discs remain completely unspoiled.

The wear widths in the group of BT-thickened greases exhibited in figure 5.7 reveal another conspicuousness. The wear intensity represented in the widths of wear marks in the sapphire balls shows a quite consistent dependence on the formulated base oil dividing this group into two almost equal subcategories. One of these is constituted by all BT-greases based on PAO and OCT-ester. The other one consists of all BT-greases formulated with HOSO and TMPO-esters. Each category resulted in similar wear widths, which differ form those of the other one. In comparison the subcategory of PAO and OCT resulted in wider marks in the balls than the subgroup of HOSO and TMPO greases. This division remains consistent throughout all applied normal forces and in all wear marks in the sapphire balls. It is not found in the widths of wear tracks produced in the discs.

Figure 5.9 displays the wear responses of an exemplary tribosystem consisting of a sapphire-ball on steel-disc material combination, which contained BT-HOSO grease as lubricant. Focusing on the wear reaction of the steel discs reveals different types of wear mechanisms, each one predominating with different intensities under different normal load conditions. On the one hand, there is a wear track of uninterrupted plastic deformation on the discs in the central region of the ball contact with some signs of surface delamination. This wear mark increases in width along with the applied normal load and is not found for 1 mN. On the other hand, there is a band of speckles surrounding the just-described uninterrupted track of plastic deformation. These speckles are even found under normal load condition of 1 mN. Its sizes increase with the normal load and they completely combine with the plastic deformation in case of 500 mN. The widths of this band of speckle patterns seems to be constant for all applied normal loads. Focusing on the wear response of the sapphire balls in figure 5.9 mainly discloses two different wear mechanisms. The images of the sapphire balls tested in all applied normal loads show abrasive wear. The sapphire balls with applied normal loads 10, 100 and 500 mN even show a typical bean-shape, which is repeatedly observed exclusively with the use of BT-greases. In some parts, this characteristic shape is accompanied by additional adhesive components. This bean-shape consists of two deeper grooves of most intensive abrasive wear of a depth of up to 166 nm and a distance of up to 24 µm leaving the main contact region in the center less affected. Moreover the sapphire balls of the three aforementioned normal loads show quite similar wear intensities. This leads to the presumption of a critical load of 10 mN for the occurrence of this mechanism in the BT-grease lubricated contact. The above-mentioned investigation of the width of wear tracks seems to corroborate this theory because all wear widths in sapphire balls created by applying 10 mN and more, in combination with BT-greases, show similar values. To better elucidate the wear mechanisms in the sapphire balls a BT-PAO was investigated under equal conditions with 500 mN of normal load. The wear marks were investigated using white light interferometry as depicted in figure A.2. This figure also shows a perspective view of the main contacting area in the sapphire ball in addition to two cross-sectioning scanning profiles in the aforementioned directions. The profile orthogonal to the direction of relative motion represented by a black solid line clearly highlights two deep grooves at the outer boundaries of contact. The profile parallel to the direction of motion reveals a peak located in the center of the contact. This peak points to abrasive processes in the inlet of the lubricant flow as well as the outlet.

Ca-greases in sapphire/steel contact The diagram of wear widths presented in figure 5.7 also exhibits the quite unique wear behavior of the group of Ca-greases. This figure makes clear that at normal forces of 100 mN and less Calcium greases produce the least if any wear in sapphire balls compared with all other greases presented in the mentioned figure. Ca-TMPO is the only Ca-grease formulation that occasions wear in the sapphire balls at normal loads of 100 mN and less. Most interestingly, and in contrast to the groups of BT and HDS-thickened greases, the Ca-grease still evoke wear in the steel discs without any detectable wear marks in the sapphire balls within the mentioned normal load region. Compared with the Li-greases and the pure base oils at normal force of 100 mN calcium greases produce lowest values ranging from 19 µm in case of Ca-OCT to 29 µm for Ca-TMPO. At a normal forces of 10 mN, Calcium greases together with pure base oils produce least wear in discs with values ranging from 8 µm in case of Ca-HOSO to 14 µm for pure HOSO base oil. In contrast to the strong base oil influences detected in the groups of clay-thickened greases, which led to the distinct division into subcategories, the wear responses evoked by Ca-greases do not appear to be subjected to base oil influences. The micrographs of an exemplary tribocontact presented in figure 5.10 with Ca-TMPO as lubricant reveals several wear mechanisms. Plastic deformation predominates the wear marks in the discs. This is detected by the relatively high flanks protruding from the polished base surface as can be seen in the normal load range covered by 500 and 100 mN. In this normal load range there are also traces of surface delamination processes discernible by dark spots within the tracks of

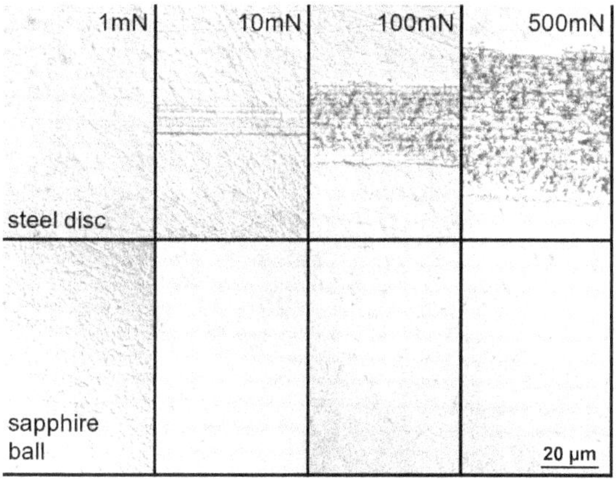

Figure 5.10.: Ca-TMPO in material combination sapphire ball on steel disc

plastic deformation in the steel plates. With 10 mN and less sole tracks of plastic deformation processes are left in the surface of the steel plates. The surface of the sapphire balls reveals a high portion of adhesive wear mechanisms that took place in the tribological stress evoked by a normal load of 500 and 100 mN. This is made manifest by metallic particles adhering to asperities of the sapphire ball surface. These particles originate from the steel surface of the revolving disc. The widths of these regions of adhesive marks on the balls corresponds to the respective tracks of plastic deformation in the steel plates. In between the metallic particles adhering to the sapphire surface there are slight traces of abrasive wear. As could have been expected from the results of wear widths presented in the bar diagram of figrue 5.7 the surfaces of the sapphire balls stressed with normal loads of 10 mN and less look completely virginal. For further investigation of the adhesive wear mechanism in combination with Ca greases and sapphire balls one of the evaluated tests was repeated with Ca-TMPO and scanned interferometrically as displayed in figure A.3 on page 156. This figure, too, exhibits a perspective view of the main contacting area and two cross-sectioning surface scans in the mentioned directions. The adhesives spots on the sapphire surface appear quite clearly in the perspective view. Moreover, the profile curves give further information on their nature and dimensions. The profile parallel to the direction of relative motion colored as a black solid line was placed outside the central region to a point where it crosses through one of the biggest adhering particles. The outcome of this profile evaluation reveals a hight of 180 nm and a width of 60 nm for this specific particle.

Li-greases in sapphire/steel contact The test results of wear widths of Li-greases presented in the bar diagram of figure 5.7 show that values of track widths in the sapphire balls generated with normal loads of 500 and 100 mN quite much resemble those achieved with pure base oils. This is also the chase with the application of a normal load of 10 mN, except for Li-PAO and -TMPO that did not generate any detectable wear, whereas the pure base oils still did. The wear results in the discs of these two groups resemble even better – with the only exception in the normal load situation of 1 mN, in which Li-greases created higher wear rates in the steel discs (8.3 µm) than the pure base oils (3.5 µm). Results also show that for normal loads of 500 and 100 mN, each width of ball wear approximates to the corresponding wear in the discs with values extending from 38 to 51 µm in 500 mN tests and from 24 to 32 µm at 100 mN of normal load. Just as was found with the Ca-greases, neither the Li-greases show the strong base oil dependence, which led to the distinct division of BT- and HDS-greases into subcategories. And neither Li-greases seem to be subjected to any base oil influences. The micrographs in figure 5.11 representing

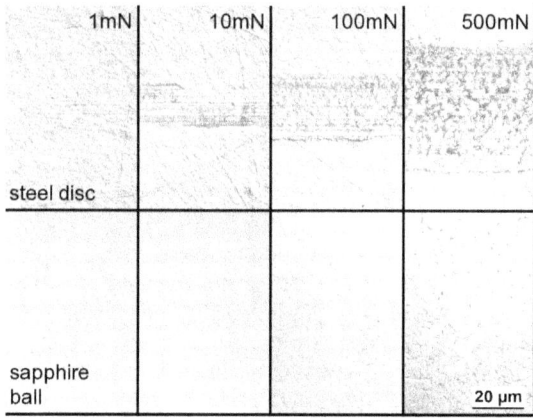

Figure 5.11.: Li-OCT in material combination sapphire ball on steel disc

the whole group of Li-greases also reveal several wear mechanisms. At first sight, it becomes clear that the wear results presented in this figure quite much resemble those achieved with Ca-greases. Here too, plastic deformation predominates the wear marks in the discs at all normal load situations. In the normal loads of 500 and 100 mN there are also traces of surface delamination processes in the steel plates as were found in the systems lubricated with Ca-greases. Here too, the sapphire balls reveal a high portion of adhesive wear in the normal loads of 500 and 100 mN and even in 10 mN. There are also slight traces of abrasive wear in the sapphire balls tested under high normal load conditions. The surface of the sapphire ball tested under normal load conditions 1 mN also appears unspoiled. As was performed with all other greases in sapphire/steel contact a selected tribosystem with Li greases was further investigated with white light interferometrical measurements too. The result is depicted in figure A.4 on page 157, where again a spacial view is presented along with surface profiles. Both of which clearly highlight the adhering metal particles in the sapphire surface. Exact evaluation reveals dimensions of 70 nm in hight and 3.5 µm in width.

Pure base oils in sapphire/steel contact The outcomes of tests with pure base oils and applied normal loads of 500, 100 and 10 mN as presented in figure 5.7 reveal that, just like within the group of Li-greases, the widths of all marks of ball wear resemble the respective dimensions of disc wear. with values stretching from 33 to 59 µm at a load of 500 mN and from 19 to 48 µm at normal forces of 100 mN. It also shows that with the applied load of 1 mN in the group of pure base oils there is only disc wear revealing values of the same sequence as with higher loads, which is HOSO, TMPO, PAO and OCT from the highest to the lowest. This sequential order leads to the clear distinction of different subcategories that was found for the groups of clay thickened greases. The only difference is that within the group of pure base oils, this sequential order is reversed, now resulting in widest wear marks for the subcategory of HOSO/TMPO and narrowest wear marks for PAO/OCT. The micrographs of selected wear results with the use of all pure base oils presented

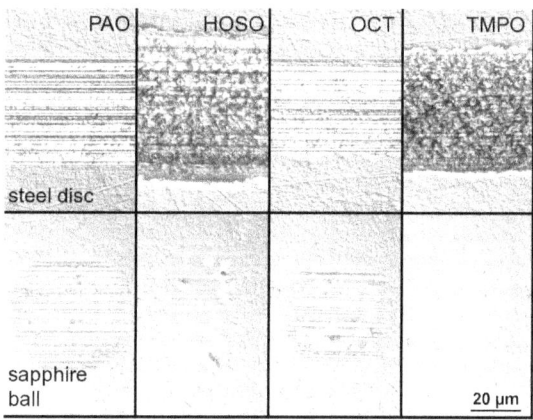

Figure 5.12.: Pure base oils in material combination sapphire ball on steel disc

in figure 5.12 exhibit clear differences between prevailing wear mechanisms in each base oil. As could have been expected by the evaluation of wear widths there is a clear distinction into the two aforementioned subcategories noticeable in these micrographs. HOSO and TMPO both resulted in mainly plastic deformational wear mechanism in the steel plates with evident signs of surface delamination. With the same oils the triboprocess evoked abrasive wear in addition to adhesive wear in the respective sapphire balls as counterparts. On the other hand, PAO and OCT both caused wear mechanisms of mainly abrasive character in the steel plates as well as the sapphire balls. In case of OCT-ester this abrasive mechanism is accompanied by adhesion processes.

5.3.1.3. Intermediate conclusions

While highly dispersed silica acid gel and bentonite thickeners most substantially affected the wear behavior toward a highly abrasive nature, lithium and calcium

5. Experimental analysis

thickeners mostly resulted in a combination of plastic deformation with slight surface delamination in the steel plates and adhesive wear in the sapphire balls. While base oil influences retain a high impact with the use of HDS and BT as thickeners, they are limited in the group of greases thickened with of Ca- and Li-soaps. Synthetic PAO and biogenic OCT ester measured as pure oils maximized the abrasive nature of prevailing wear mechanisms but they minimized the widths of the measured wear tracks in sapphire balls and steel discs. On the other hand they also maximized the abrasive behavior of all clay thickened grease systems. HOSO and TMPO ester, which mostly generated plastic deformational wear mechanisms resulting in wider wear marks when measured as pure oils, minimized the abrasive behavior of the clay greases. They also favored the effects of adhesion of particles to the sapphire ball surface and the intensity of surface delamination in the surface of the steel discs. Ca-thickened grease formulations generally resulted in least wear and lower friction coefficients. Lithium greases resulted in values for the coefficient of friction and wear widths similar to base oils although wear depths and wear mechanisms were much different.

5.3.2. Metal soap-, HDS- and BT-greases in steel-steel contact

All of the tests performed in the previous section were repeated with material combination steel ball on steel disc. In this given material combination the preset normal loads resulted in values of Hertzian stress as shown in table 5.3.

Table 5.3.: Normal forces and resulting values of Hertzian stress in the nanotribometer—steel balls and steel discs

normal force [mN]	500	100	10	1
max. Hertzian stress [GPa]	1.31	0.77	0.36	0.17

5.3.2.1. Friction results

Frictional results in material combination steel ball on steel disc are exhibited in figure 5.13 in logarithmic scale. Just like in the sapphire/steel contact situation the frictional values in the steel/steel contact also show a clear dependence on the applied normal load, which can be attributed to the aforementioned rheological resistance of grease films in rotatory tribological tests. Again, the relatively low coefficients of friction produced in normal load situations with 10 and 1 mN and contacts lubricated with pure base oils, which because of low viscosity do not offer a rheological resistance, corroborate this theory. For this reason, and because of higher fluctuations in the low normal load region, the ensuing evaluation will again be restricted to the high normal loads 500 and 100 mN.

In the evaluation of presented results from the thickeners point of view one first of all finds that the differences between the groups are smaller in steel/steel contact than in sapphire/steel contact. Still, there are slightly higher values of coefficients of friction detectable in the group of HDS-thickened greases (0.091 ± 0.012) than for

5.3. Tribological analysis

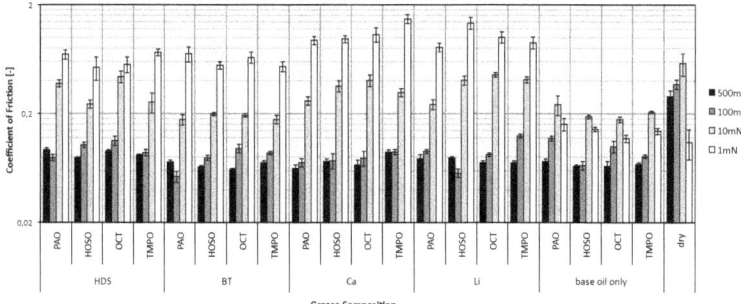

Figure 5.13.: Friction in contact situation steel ball on steel disc—all greases, pure base oils and dry contact situation

all other groups—only dry contact aroused higher values. The ranking continues with Li (0.082 ± 0.020), Ca (0.076 ± 0.009) and BT (0.073 ± 0.014).

When comparing the determined values of friction coefficient from the base oils' point of view one finds no tendencies strong enough to be mentioned.

5.3.2.2. Wear results

Widths of all wear tracks in the steel balls and the steel discs were microscopically determined and compared with each other in figure 5.14 in the known manner, with wear dimensions in the steel discs negated for a better comparability. In this image the same scale was used as in the sapphire/steel contact. This fact results in missing presentability of the values gained in dry contact situation with its data outside the displayable range. Again, wear responses will be discussed along with representative

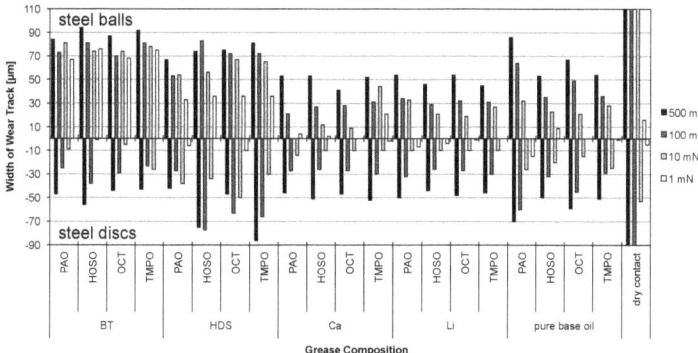

Figure 5.14.: Widths of wear tracks in steel balls and steel discs—all greases, pure base oils and dry contact

85

5. Experimental analysis

wear marks provided by micrographs displayed in figures 5.15 to 5.18.

HDS-greases in steel/steel contact The interpretation of the data of wear widths in steel/steel contact situation presented in figure 5.14 from the thickeners' point of view makes clear that the HDS-greases, following right behind the BT-greases, evoke the second widest wear marks in the steel balls compared to all other greases and to the pure base oils. This figure also depicts that widths of ball wear quite well correspond to the respective widths of disc wear. This correspondence is quite persistent throughout all applied normal loads except for HDS-HOSO and -TMPO at 1 mN of normal load. In fact, the wear widths measured in the discs are the widest in all lubricated systems in the whole series of experiments with values up to 86 µm in case of HDS-TMPO. Additionally, this mentioned correspondence between ball and disc wear is most distinct within the group of HDS greases compared to all other greases and is only found as distinctly in the group of pure base oils.

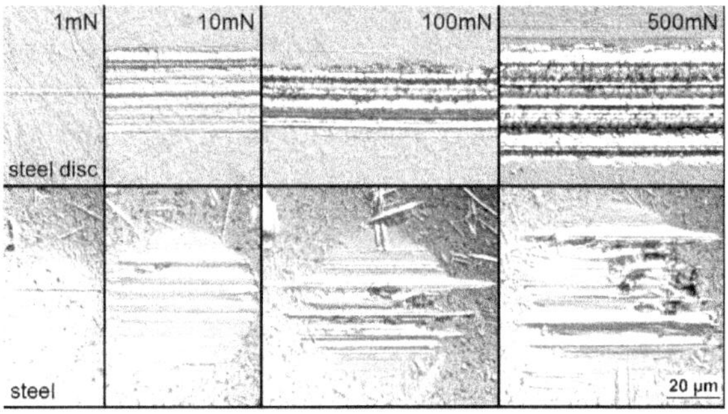

Figure 5.15.: HDS-PAO in material combination steel ball on steel disc

The highly abrasive character of HDS-thickened greases in steel/steel contact is depicted very clearly in figure 5.15. Abrasion appears to be the main wear mechanism in both the steel plates and the steel balls. In the normal load region of 500 and 100 mN this abrasive mechanism is accompanied by a slight proportion of surface delamination in the steel discs as can be seen by the dark spots representing cracks orthogonal to the direction of movement. The property of HDS-greases resulting in polishing of the surface of the balls while leaving contacted parts of the discs unaffected, as mentioned in the evaluation of sapphire/steel contacts, is also observed in steel/steel contact, especially when applying 1 mN. This polishing steel/steel contact, however, is not as smooth as in the sapphire/steel contact. The images of disc wear also reveal that elevations created in the steel discs evoked grooves in the respective balls.

BT-greases in steel/steel contact The first conspicuousness in the diagram of wear withs exhibited in figure 5.14 are the almost consistent high values of wear widths in steel balls in combination with the group of BT-greases. They very clearly tower above all other groups of greases sorted by individual thickeners. These mark widths are consistent throughout all normal loads applied during the experiments. Even in 1 mN, the least of all applied normal forces, they reach 71,5±4.65 µm and lie far beyond all widths of wear marks created with 500 mN in steel balls with the investigated metal-soap thickened greases. The evaluated data presented in figure 5.14 depicts no correspondence between ball and disc wear. When interpreting the present data within the group of BT-greases from the base oils' point of of view one finds the aforementioned division into subcategories PAO/OCT and HOSO/TMPO. Yet the change of material combination reversed this effect unto slightly higher resulting values for the subgroup of HOSO/TMPO.

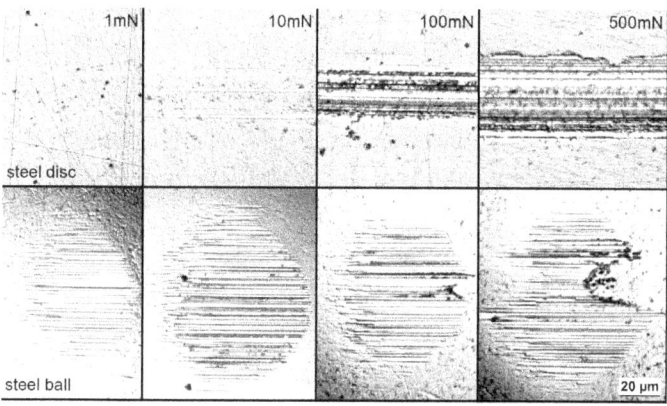

Figure 5.16.: BT-TMPO in material combination steel ball on steel disc

Figure 5.16 displays the wear results of a tribosystem lubricated with BT-TMPO in a steel-ball on steel-disc material combination. Analysis of the wear marks produced in the steel plates discloses several wear mechanisms. Firstly, wear marks of mainly abrasive nature in the central contacting area of the steel plate surface can be detected. These marks are most clearly visible in normal load regimes of 500 and 100 mN. While the image corresponding to 10 mN still reveals slight abrasive lines, there are no such wear tracks by applying 1 mN normal load. Secondly and just like in sapphire/steel contact, there is a band of evenly distributed speckles to be detected around the central contacting area. The circular wear marks in the steel balls also reveal a highly abrasive wear mechanism with wear grooves of an average width of 1.26 µm for all normal loads. The characteristically waisted wear mark shape produced in a normal load regime of 500 mN is found for all BT-greases tested.

Ca-greases in steel/steel contact The diagram of wear widths displayed in figure 5.14 reveals that the group of Ca-greases, just as it did in the sapphire/steel combination, results in the least wear compared with all other tests in steel/steel contact with values decreasing significantly along with the applied normal load with Ca-TMPO being the only grease in the group of Ca-thickened ones, which generates

5. Experimental analysis

wear in the discs with 1 mN of normal load. The wear dimensions in the balls quite much resemble those achieved in the discs. Influences of the base oils in the Ca-thickened greases are too small to be detected or even interpreted into tendencies. Only the results determined at a normal load situation of 500 mN would allow to be interpreted into trends of such a behavior which then resulted in the formation of the two known subcategories PAO/OCT and HOSO/TMPO. Micrographs of wear

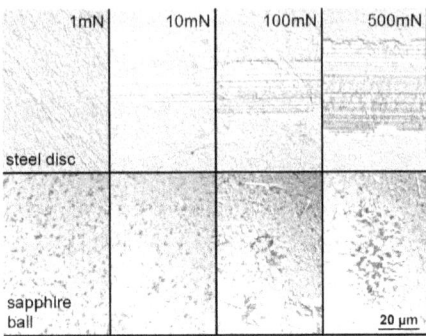

Figure 5.17.: Ca-PAO in material combination steel ball on steel disc

marks created in the discs with the use of Ca-greases as presented in figure 5.17 reveal that wear processes within this group are predominated by mechanisms of plastic deformation with only very slight traces of abrasion. This two-fold influence is found in the high normal load conditions of 500 and 100 mN. By applying 10 mN normal force, there are only traces of abrasive mechanisms and non-detectable wear at the application of 1 mN normal load. The wear marks in the steel balls reveal adhesive wear mechanisms in those systems tested with a normal load of 500 and 100 mN. The steel surfaces of the balls tested with 10 and 1 mN appear unspoiled.

Li-greases in steel/steel contact Generally, the soap greases evoke wear responses of least intensity in all of the investigated lubricants. The group of Li-greases presented in figure 5.14 is noticeable also with comparatively low widths of wear marks. The results measured in the normal load situation of 500 mN in fact quite much resemble the results determined with Ca-greases. Only the results gained from tests under low normal load condition 10 mN differ from those achieved with Ca-greases inasmuch as the wear widths in the steel balls are a little wider. In the normal load of 1 mN none of the Li-greases resulted in wear marks in the steel balls. Pertaining the widths of wear marks in the steel discs, the Li-greases also resemble the Ca-greases with the only difference of still detectable wear under normal load condition of 1 mN. In the group of Li-greases the dimensions of wear marks in the balls correspond quite well to the respective wear marks in the discs. In Li-greases the division in to the two known subcategories PAO/OCT and HOSO/TMPO can only vaguely be detected in the results originating from higher normal loads of 500 and 100 mN. In this case it results in higher values for the category of PAO/OCT than for HOSO/TMPO.

5.3. Tribological analysis

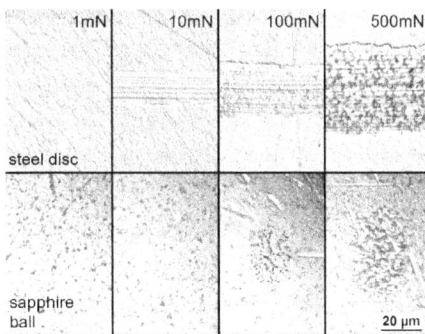

Figure 5.18.: Li-HOSO in material combination steel ball on steel disc

The micrographs of wear marks of a Li-HOSO grease in steel/steel contact situation as a representative tribosystem displayed in figure 5.18 reveal prevailing wear mechanisms quite well. The normal load range from 500 down to 10 mN clearly depicts plastic deformation as the main wear mechanism in the steel discs. When applying a 500 mN normal load there is also surface delamination made manifest by black areas orthogonal to the direction of relative motion in the track of plastically deformed steel disc surface. The load of 1 mN left the surface of the steel plate unaffected. The wear marks in the steel balls stressed with a normal load of 500 and 100 mN reveal a sole adhesive wear mechanism, detectable by dark spots arranged ovally in the main contacting area. On the other hand the micrograph taken from the tribosystem that was stressed with 10 mN exhibits some particles of adhesive wear along with clear marks of abrasive wear.

Pure base oils in steel/steel contact The values presented for the group of pure base oils in the diagram of widths of wear tracks in the steel balls and steel discs in figure 5.14 exhibit a quite discernible behavior for each base oil. The very good correlation between ball and disc wear in each respective tribological pairing presents itself very clearly in this figure. Slight deviations from this behavior are only detectable in the normal load condition of 1 mN. The fact that this behavior leads to a clear division of the base oils into the known subgroups of PAO/OCT and HOSO/TMPO, in both the disc and the ball wear responses, is very conspicuous. It results in much higher values for PAO/OCT than for HOSO/TMPO. The only other group that resulted in such a clear division into the aforementioned subcategories in both, the steel discs and the steel balls, is the group of HDS greases. Generally the group of pure base oils resulted in slightly higher dimensions of wear widths than the groups of soap thickened greases.

The micrographs of representative signs of wear determined under a normal load condition of 500 mN and with the use of pure base oils as displayed in figure 5.19 also show this well discernible behavior of the two subcategories of PAO/OCT and HOSO/TMPO by arousing individual wear mechanisms in each subgroup. The latter evoked mainly plastic deformation in the steel discs with first signs of surface

89

5. Experimental analysis

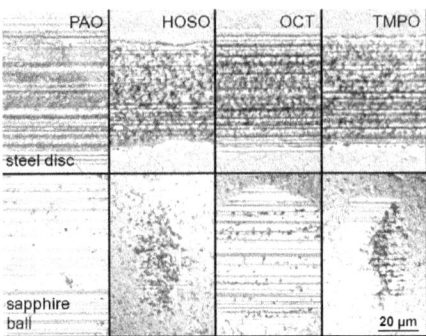

Figure 5.19.: Pure base oils in material combination steel ball on steel disc

delamination and abrasive wear. It also caused adhesive wear in the steel balls with traces of abrasive wear mechanisms, which are much stronger prevailing in the tribocontact lubricated with TMPO than in the one lubricated with HOSO. The sub group of PAO/OCT on the other hand, resulted almost exclusively in abrasive wear mechanisms in both the steel balls and the steel discs. The wear mechanism of surface delamination is also present in the tribocontact lubricated with the OCT-ester.

5.3.2.3. Intermediate conclusions

Wear results created with tribosystems consisting of a steel ball on steel plate contact with greases containing traditional clay and metal soap thickeners and biodegradable base oils were presented and discussed in this section. Results of micrographic wear analysis revealed more intensive wear with the use of clay thickeners than with the use of soap thickeners. While clay thickeners mostly evoked abrasive wear mechanisms, the soap thickeners generated wear mechanisms of plastic deformation and slight surface delamination. Comparison of single results within all groups of thickeners showed a dependence on the formulated base oils. Comparison of all thickener groups to wear tests with pure base oils delineated similar influences in the groups of soap thickened greases and contrary influences within all clay thickened greases. With the exception of BT-greases, all thickener groups resulted in widths of ball wear that resemble the respective widths of wear in the discs.

5.3.3. Comparison of tribological results with different material combinations

All results discussed in sections 5.3.1 and 5.3.2 clearly show several effects that demand an ensuing elucidation. For one thing, there are effects, which can be distinctively associated with the applied base oil, the formulated thickener and the material combination during the tests. For another thing the highly abrasive character of all clay grease systems was detected along with a "speckle forming"

5.3. Tribological analysis

process found with the use of BT-greases. All these effects are discussed in this present section.

5.3.3.1. Base oil-, thickener type- and material combination-dependent frictional and wear responses

Several very interesting effects come to light when comparing the so far presented results from the point of view of material combination. This comparison is performed in the present section along with a discussion of explanations for the found effects. Most observations refer to the diagrams of wear widths created with all investigated greases as shown in figures 5.7 (sapphire/steel contact) and 5.14 (steel/steel contact) in the previous sections.

The analysis of wear results from the material combination point of view reveals different system responses depending on the thickener. Comparing the groups of clay thickened greases, the mentioned diagrams of wear widths clearly show that the group of BT-greases in sapphire/steel combination resulted in much less ball-wear than HDS-greases did. Wear widths found with HDS-greases in some parts even amount to twice the value of those achieved with BT-greases for the higher applied normal forces (500 and 100 mN). With lower normal loads (10 and 1 mN), however, wear marks created with HDS-greases are only around 1.4 to 1.5 fold wider than the ones produced with BT-greases. Besides this, while in sapphire/steel contact BT-greases resulted in the narrowest wear marks in the balls, in all of the examined clay greases, with an average value of $37.81 \pm 9.51\,\mu m$, they produced very wide ball wear scars ($78.44 \pm 7.95\,\mu m$) in steel/steel contact. Most interestingly, this effect is slightly reversed for the group of HDS-greases with very wide wear marks in the sapphire balls ($63.38 \pm 23.41\,\mu m$) but relatively narrow ones in the steel balls ($60.00 \pm 17.05\,\mu m$).

Another conspicuousness that becomes apparent when assessing the results presented in the just mentioned figures 5.7 and 5.14 from the material's point of view relates to the interaction of material combination and the base oils. On the one hand, in all sapphire/steel contacts, PAO and OCT-ester in combination with both clay thickener types resulted in the widest wear marks produced on sapphire balls within their respective thickener group. In the group of HDS-greases the impact of this effect is most severe producing wear marks 57.1% wider than the ones resulting from the use of HOSO and TMPO-ester. While not as severe, this effect is still obviously apparent within the group of BT-greases, resulting in 28.2% wider wear marks (given numbers were averaged for all normal forces within the respective thickener group and depend on and generally diminish along with the applied normal load). On the other hand, the influence of the base oil on the widths of all ball wear marks is reversed in the steel-ball on steel-disc material combination. Here, the subcategory of HOSO/TMPO resulted in the widest wear marks (8.0% wider within the group of BT-greases and 9.4% wider in the group of HDS-greases) compared to those obtained with PAO and OCT.

The assessment of the present data also makes clear that while the wear marks in the specimen balls of clay thickened greases revealed the above-described clear distinction between BT- and HDS-greases or the different base oils, the widths

5. Experimental analysis

of wear marks in the discs do not all show this same clear response. In fact, the wear widths in the discs of the greases tested in sapphire/steel contact quite much resemble; the only difference is that the wear widths produced with HDS-greases are 14.8% wider in average. When comparing the extent of wear marks in the discs to the ones left in the sapphire-balls one finds no correlation of these two within the group of greases tested in sapphire/steel contact. The situation is different in steel/steel contact, however. Here, the same interactions predominate between formulated base oils and their respective wear widths as were found in the steel balls. While they are only to be detected as general trends and for higher normal forces within the group of BT-greases, they are clearly present for all HDS-greases resulting in widest wear marks when the contact is lubricated with HOSO- and TMPO-based greases.

Wear results in steel/steel contact also make clear that base oils on their own behave differently in the tribocontact than base oils in formulation with different thickeners. The OCT ester and PAO measured in 500 and 100 mN normal load regime are the oils with the highest rates of abrasive wear in the steel balls and the discs, as pictured in figure 5.19, and the widest wear marks as can be seen in figure 5.14. This characteristic is also apparent within the groups of soap-thickened greases, except its impact is not as strong and only reliably detectable in the high normal load condition of 500 mN. The groups of clay-thickened greases on the other hand, turn this effect to the opposite with the subcategory of HOSO/TMPO now resulting in the highest wear widths in the steel balls. Pertaining the disc wear only the group of HDS greases shows this behavior.

This same effect is also detected in the sapphire/steel material combination. Here too, wear results make clear that pure base oils behave differently on their own than in combination with a thickener. Most interestingly, the effects described in the last paragraph are completely reversed to the opposite with the subcategory of pure base oils PAO/OCT now generating least wear widths in the sapphire balls and discs. Notwithstanding the less wear widths they still generate the highest abrasive wear mechanism compared with the group of HOSO/TMPO as depicted in figure 5.12. Now these characteristics are not copied by the groups of soap thickened greases as clearly as it happened in the steel/steel contact. The groups of clay thickened greases again reverse the base oil influence to the opposite, although this effect is only apparent in the sapphire balls with the subcategory PAO/OCT now generating the highest wear intensities.

All of these effects, which describe the dependence of wear responses on the kind of base oil, the used thickener type and the tested material combination are presumed to be attributed to the polarities of the selected base oils as has been predicted and modeled in the assumptions in section 3.1.1. It is assumed that the base oil polarity highly influences the interaction of the whole base-oil-thickener-surface system. The polarity of the selected oils was measured and discussed in section 5.1 by means of interfacial tension measurements. Results are depicted there along with the polarities of the tribological surfaces. Yet, one must bare in mind that the prevailing main types of van-der-Waals interactions may change in the course of a test, especially taking into account the abrasion of highly polar oxide layers in metal surfaces. In order to elucidate the base oil polarities' influence on tribological interactions of

5.3. Tribological analysis

the investigated surfaces the above-discussed results of tests with pure base oils presented in figure 5.7 and figure 5.14 are highlighted again. The results very clearly indicate the presence of a protective layer, which is formed due to the high polarity of HOSO and TMPO and resulting Keesom-bonds in physical adsorption processes. These are made manifest by the comparably narrow wear scars with the application of the respective oils in steel/steel combination and in both contacting partners. On the other hand, the lower polarity oils, PAO and OCT, resulted in wider wear marks and thus offered less protection. This effect is reversed, however, in sapphire/steel contact—here the high-polarity oils offered least protection if judged by means of sole wear widths, which might be ascribed to the smaller relative polarity of sapphire and to poorer wettability behavior in the combination of sapphire solid surfaces and highly polar oils. Regarded by the abrasiveness of the wear character as depicted in figure 5.12, however, the least protection might be interpreted to the category of PAO/OCT. When applying these deductions to the interpretation of tribological investigations of the given clay thickened grease systems one finds a perfect reversal of polarity influences in comparison of wear results of the pure oils to those achieved with the respective greases (figure 5.7). When considering the reasons for this effect one must keep in mind which mechanisms may work within the greases and between the surfaces. Several, partially opposing, mechanisms come into effect with the use of high-polarity oils in combination with clay thickeners and differing material combinations.

For one thing, there is the influence of the thickening process: high-polarity oils, more than low-polarity ones, tend to deposit to outstanding silanol groups in primary particles of HDS-greases and to hydroxyl groups of organosilicon particles in BT-greases, thus interfering with and partially preventing the formation of a three-dimensional network and causing a higher need of share of thickener content for equal mechanical stability of the clay-grease as shown in table 4.2. In this table it is evident that more thickener is needed in combination with highly polar oils than with low-polarity oils. Moreover, the influence of this effect is bigger in BT-grease systems than in HDS-greases. This may be substantiated by the use of quarternary ammonium ions for organophillic modification of the BT-greases, which were kept the same for all given BT-greases in the process of formulation. This resulted in similar hydrocarbon chain rest molecules, which in turn deeply influenced the bonding character between thickeners and oils of different polarities. For further investigation of this subject it is proposed to use tailor-made quarternary ammonium salts to fit the polarity of each base oil. This way the share of thickener content would be kept at a fixed level and the direct influences of base oil polarities could be investigated without interference of thickener content. Also for further investigation, in this context, it is proposed to examine the rheological grease behavior depending on base oil polarity. In conclusion, this effect leads to the creation of a physisorptive protective layer of highly polar oil surrounding the extremely hard SiO_2-compounds of both clay grease types. This may explain the comparably narrower wear tracks in the sapphire balls in combination with high-polarity greases BT-HOSO and -TMPO even though more of these hard particles are present.

For another thing, there is the influence of physisorptive adhesion processes in association with highly polar oils and metallic tribological surfaces, which leads to less wear with the use of pure oils as described above. In combination with clay thickeners, however, metal oxide layers of tribological surfaces seem to more attract highly-polar oils than clay-particles and thus leading to a destruction of the just-described protective layer surrounding them. This, however, leads to a higher number of unshielded extremely hard clay particles in the tribological steel/steel contact resulting in higher wear rates for clay greases with highly polar oils than with low-polarity ones.

Besides this, the group of BT-greases provided lower friction coefficient values compared to HDS thickened greases, independently of material combination and normal load. The only exception to this rule is BT-TMPO within the normal load regime of 1 mN—in both material combinations it resulted in higher coefficients of friction compared to HDS-TMPO. Because of the very low normal load of 1 mN and its resulting rheological significance, explanations for this exception are presumed to be found rheologically. For this reason, further rheological grease investigation is suggested. Also, in both material combinations, the use of different base oils did not influence the frictional results in the group of BT-greases as significantly as it did within the group of HDS-greases. In this group, HDS-PAO- and HDS-OCT-based greases generally led to higher values of the friction coefficient than HDS-HOSO- and HDS-TMPO-based greases, for all applied normal loads. This difference is even more intensified with a change to a sapphire/steel material combination, generally resulting in higher frictional values when applying HDS-greases. The examined BT-greases, however showed a much more stable frictional response throughout both material couplings and no clear influences of both the thickeners and the base oils.

The quite distinctive frictional and wear behavior in combination with all investigated soap-thickened greases can be elucidated only when taking the microphysical mode of action of lubrication into consideration. Lubrication only comes into effect by the application of a lubricant film that mechanically separates the contact partners [24, 75]. Both, base oil and thickening agent influence the formation of this separating film in lubricating grease [34, 130]. The results of performed series of experiments show that the thickener's influence is more substantial in this context. To some extend explanations for these effects can be found in the microstructure of the thickening agents. Metallic soaps usually form three-dimensional networks [94] of polymorphic soap fibers that confine the oil with secondary valence bonds. It is well known, and has been shown in section 5.2, that lithium-greases generate longer and more densely arranged and entangled fibers than calcium-greases [36, 41, 123, 131, 132]. This difference could be an explanation for the lower friction coefficient and the generally lower wear rates of examined calcium-greases compared to the greases formulated with lithium-soap.

Microstructural analysis of some exemplary Li-greases by means of atomic force microscopy, as performed in section 5.2, helps elucidate these effects. AFM images displayed therein have shown a direct influence of the base oil on the formulation of soap fibers. In combination with HOSO and TMPO the Li-soap had resulted in long, mostly parallel and less entangled fibers as shown in section 5.2, whereas in combination with OCT it had resulted in rather short and more entangled fibers. As explained earlier in this work, it is assumed that shorter soap fibers result in less friction and less harmful wear mechanisms. In deduction it is assumed that the less abrasive character of shorter soap fibers in general and specifically in combination of Li-soap-thickener and OCT base oil counteracts the higher abrasive properties of OCT, resulting in less to equal wear intensities and frictional values.

5.3.3.2. Creation of "speckle patterns" with the use of BT-greases

AFM investigations of the worn surfaces provide more information about the nature of these very typical speckles as shown in figure 5.20. They seem to consist of elevations with very nearby wholes in dimensions depending on the age of the speckle.

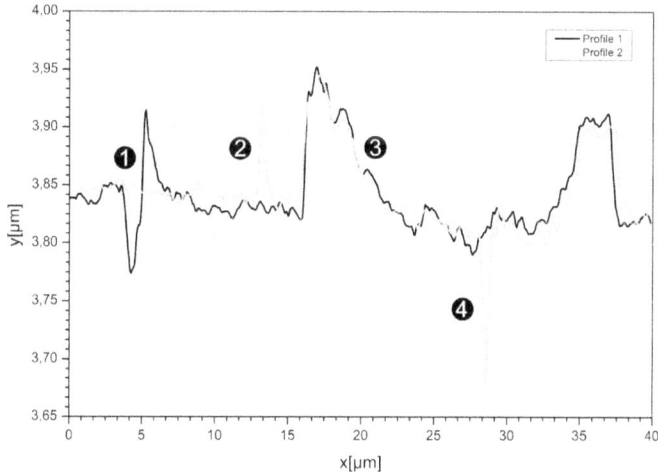

Figure 5.20.: AFM Profile of speckles and wear track—BT-grease 100 mN in material combination sapphire ball on steel disc—Note: Points of emphasis: 1)Older speckle in distance of 22.5 µm from the center of the main groove (depth 74 nm, height 80 nm). 2)Younger speckle in distance of 13.9 µm from the center of the main groove (height 84 nm. 3)Flank of main groove (height 108 nm, total penetration depth 52 nm). 4)Older speckle in the center or the main groove (depth 176 nm).

Two line profiles are depicted in figure 5.20. They go through a main deformational wear track (3) with one adjacent speckle each (1) and (2). The one that is referred to as point of emphasis number (1) presents itself as a whole of 74 nm in depth with an adjoining elevation of 80 nm in height. The speckle that is flagged in point number (2) is exclusively structured as elevation (height 84 nm). The formation process of these speckles is assumed to be able to commence at any time in the course of the experiment with the interlinking of small montmorillonite particles of the bentonite thickener to the surface of the steel plates. The physically working adhesive mechanisms will be discussed later. The fact that the corresponding sapphire counter surface wears off in most parts of the regions where these speckles

5. Experimental analysis

occur indicates to extreme hardness of the speckles. This presumption becomes even more profound regarding the fact that the steel surface around the speckles appears completely unspoiled although the sapphire is much harder than the steel. In further progression of this process it seems plausible that other particles are more likely to adhere to these protruding elements and thus allowing the speckle to grow. Therefore it is deduced that these protruding particles offer a target to the friction process resulting in high shearing stress in their base, which will presumably plastically deform subjacent regions resulting in very nearby wholes. Therefore it is concluded that since there is no adjoining whole in the speckle identified in point of emphasis number (2) of figure 5.20, it is younger than the one marked as number (1). Another older speckle is indicated with point (4). It is located in the very center of the main deformational groove and by its depth (176 nm) it is presumed to be more advanced than the ones described so far. Also it is assumed that it was produced through the above-described process apart from the fact that the typical elevation was cut off as a result of higher surface pressure in the main contacting area between ball and disc in the course of experiment. An indication of this theory is the appearance of the wear mark evolved with a normal load of 500 mN in the steel disc as depicted in figure 5.9—here it shows very clearly that all speckles were consumed by the deformational wear mark.

The aforementioned disquisition on mutual interdependence of bonding mechanisms between oils, solid body surfaces and thickener surfaces, as postulated and modeled in section 3, leaves room to conclude that the same mechanisms may lead to adhesion of clay thickener particles to tribological surfaces. This may elucidate this "speckle forming" process. Moreover, it has been observed during the evaluation of wear images taken after the tests that highly polar oils generally tend to support this speckle forming. This may add another opponent to the protective character of highly polar oils in clay-thickened greases: Provided that these speckles also form on the surface of the investigated steel balls, it may be deduced that they add to the highly abrasive character of BT-greases when they come into contact with the speckles on the steel plates during the course of the experiment. Therefore, it is assumed that the speckles especially have a negative influence on the wear behavior if they occur in both contacting surfaces.

5.3.3.3. Highly abrasive behavior of clay greases

The highly abrasive character of all examined mineral thickeners might be explained by their microphysical constitution. All of the above-presented images of wear appearance indicate that the microstructure of BT- and HDS-greases differ significantly from each other and from those of soap-thickened greases. With the application of HDS-greases there were typically polished regions found in the specimen balls in both material combinations, whereas BT-greases generally led to rougher worn off caps in the balls.

As has been shown in the AFM analysis in section 5.2, HDS-OCT grease contains very small solid SiO_2 particles. It is presumed that these small and hard particles induce a polishing material removal process in the contacting specimen balls. The presence of these particles has been proven in the residue of wear debris via EDX-analysis [132,133]. The high proportion of abrasive wear with the use of HDS-greases especially in the sapphire balls is to be explained by the extreme hardness of these SiO_2 particles (5 to 7 on the hardness scale according to Mohs). Although sapphire is even harder (up to 9 on the Mohs scale) it yields the persistent influence of SiO_2 particles. It is presumed that these SiO_2 particles are elastically embedded in the metal surface of the steel plates during the course of performed frictional tests without abrasively influencing the steel. This would explain why in some parts the sapphire balls wore off while the corresponding contacted area in the steel plates was left unaffected. These results are in agreement with other studies previously reported [134], where the effect of the tribologically active SiO_2 nanoparticles was investigated. A comparison of these results with the present study suggests SiO_2 particle sizes of less than 100 nm, as may be corroborated by figure 5.5 in the AFM analysis.

The widths of typical wear grooves generated in the surfaces of steel balls with the use of BT-greases propose a similar size for BT particles in comparison with those found in the AFM analysis in section 5.2. There it had been shown that BT-HOSO grease revealed a large amount of evenly distributed solid platelets of montmorillonite particles. It had also been shown that these montmorillonite particles are substantially larger than SiO_2 particles of HDS thickener. It is presumed that the physically adsorbed aforementioned "speckles" along with elastically embedded particles of montmorillonite work abrasively in the contacting steel and sapphire balls.

5.3.3.4. Intermediate conclusions

The investigation of greases based on biodegradable base oils and mineral thickeners, bentonite and highly dispersed silica acid, in addition to metal soap thickeners, revealed a clear distinction between each main component's influence on frictional and wear behavior. Findings moreover disclose a different behavior of these components depending on the material combination. Soap thickened greases generally led to less wear in the balls and discs in both investigated material combinations and showed less dependence on the base oil influence. The clay thickened greases generated much wider wear widths and substantially more abrasive wear mechanisms. Moreover, they depicted a sensitive dependence on the base oil influence, which most interestingly worked in contrary dimensions to the measured pure base oils in both material combinations. Frictional and wear results correlate very well in sapphire/steel contact resulting in high coefficients of friction for tribo-contacts with high wear rates. In steel/steel contact, however, this effect is not found. All detected base oil dependent influences are attributed to polarity effects.

5.4. Rheological analysis

In the previous chapters, the strongly prevailing influence of base oil polarities on the tribological behavior of lubricating greases has been expounded and discussed. The physical working mechanisms of lubrication have been discussed in prior sections of this work. The preset parameters of the performed tribological tests, especially

5. Experimental analysis

the normal load and the velocity of relative motion between the surfaces in the contact situation, indicate to a state of mixed friction. Since the lubricant film, which separates the contacting partners in the mixed friction state is mostly strained in areas of least film thickness, it only seems logical and consequent to focus on the lubricating grease for further investigation of the phenomena found. For this purpose one has to realize initially, the way in which the lubricant film is generally loaded in the contact. Because all of the so far performed and discussed tribological investigations were performed in sliding friction in the state of mixed friction it is presumed that the main part of the lubricating grease partaking in the active process of lubrication was loaded by shear strain. The use of rheometers seems appropriate for simulation of this load case of shear strain in the mixed friction state. Even though the shear rates applicable in rheometers will never even come close to those prevailing in the mixed friction state of highly loaded frictional pairs they still offer a meaningful, if not the only, alternative in this endeavor. In almost all rheological investigations of lubricating greases, the response of the measuring sample to a certain load is observed and monitored. Even though these loads are much smaller than the ones prevailing in real tribocontacts, modern rheometers, with their sensitive measuring equipment, are able to detect even the smallest of changes in the sample. It is up to the user to decide if and how the results detected with a rheometer are to be extrapolated to the real contact situation. For this purpose, all of the selected greases were investigated with several rheological measurements, all described in section 4.2.4 of this work. These investigations will show if polarity-induced influences found tribologically will also predominate in rheological investigations. And, should this be the case, this fact would make it necessary to find out if tribologically found influences are to be attributed to the rheological grease behavior.

5.4.1. Rotational transient tests

5.4.1.1. Results and discussion of rotational transient tests

Figure 5.21 shows different transient flow curves for selected Li-grease samples formulated with all examined base oils. This graph quite clearly shows how well the mathematical approximation according to equation 2.38 [24] fits the real evaluation of shear stress over time in most cases and in almost the complete temporal evolution. Only the initial range of the mathematical fits do not represented the real data very well throughout all curves and especially with Li-PAO, which shows a large discrepancy between the fit and the real value plot in the beginning. This demonstrates the weakness of the proposed mathematical approximation.

In order to obtain values of rheological energy densities $e_{rheo-rot}$, as defined in equation 2.39, the monitored development of shear stress over time was numerically integrated within bounds corresponding to τ_{max} and τ_{lim}, respectively, as limiting time constraints. It is important to mention that the real measurements were integrated and not the fits according to equation 2.38. Results of these integrations of tests performed in exemplary temperature conditions of -10 °C and 40 °C are displayed in figures 5.22 and 5.23 with greases arranged in groups of thickeners. For the purpose of a better discernibility, the values were converted into mJ/mm^3 and

5.4. Rheological analysis

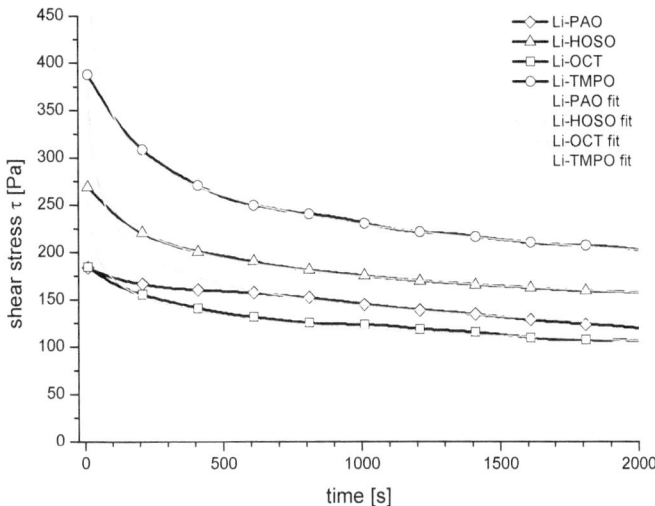

Figure 5.21.: Transient flow curves of selected Li-grease samples determined at 40 °C along with the fits of the corresponding curves according to equation 2.38 [24]

are displayed in linear scale.

The diagram in figure 5.22 clearly depicts a good correlation between energy densities and base oil polarities. All greases based on the highly polar oils, HOSO and TMPO, show a higher energy density within their respective thickener group than those based on the low-polarity oils, PAO and OCT. The only exception to this rule is the HDS-TMPO system, which exhibits a value of $e_{rheo-rot}$ lower than expected. However, this result does not constitute an outlier because it is physically explicable by some effects, which occurred only at very low temperature conditions with this grease. Thus, in all tests at -10 °C, it turned into solid matter, which was expelled from the measuring plate-plate gap at around 100 s. So the true value of energy density of HDS-TMPO, although not detected, is much higher than displayed in the diagram. The difference of energy densities between greases formulated with high-polarity and low-polarity oils, respectively, is most apparent in the group of BT-greases (72.4%), followed by Li-greases (48.7%) and Ca-greases (46.5%). If ignoring the value obtained for the HDS-TMPO system, the average difference of energy densities between high and low polarity greases rise up to 88.6% within the group of HDS-thickened greases.

The diagram of energy densities determined at 40 °C displayed in figure 5.23 does not show this distinct differentiation between greases based on high and low polarity oils for all grease systems. Only Li-thickened greases reveal such behavior. Most interestingly the effect is reversed in the BT-thickened greases, with low polarity oils PAO and OCT now evoking higher energy densities in the greases. As previously reported [99], compositional variables are the most influencing factors

5. Experimental analysis

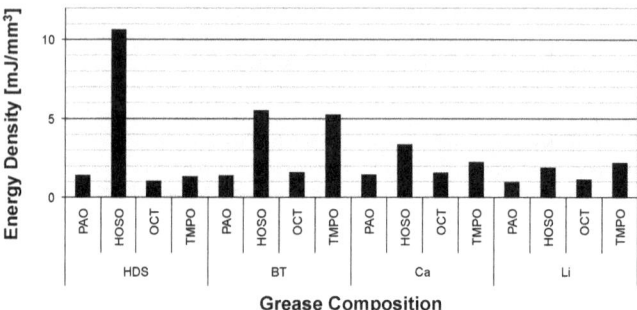

Figure 5.22.: Energy densities, $e_{rheo-rot}$, determined by integration of shear stress over time at -10 °C

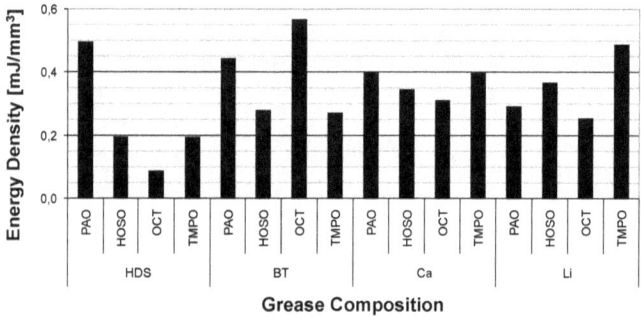

Figure 5.23.: Energy densities $e_{rheo-rot}$ determined by integration of shear stress over time at 40 °C

affecting the rheology of greases. Particularly in this investigation, the nature of the different components (base oil and thickener), with individual solid or liquid characteristics, and concentrations are considered. The base oil viscosity and especially the percentage of thickener exert an influence on the structural degradation. The base oil viscosities do not differ significantly from each other in the given grease systems. The situation is different, however, with the percentage of thickener as shown in table 4.2. These values reveal a polarity dependence of thickener percentage for all clay-thickened greases and for the Li grease systems to fit a desired NLGI class 2. This influence is made manifest by a higher thickener ratio needed for highly polar oils than for low polarity oils. Ca greases, on the other hand, do not show this dependence, probably due to the presence of the double charged cation. This influence most severely comes into effect within the groups of BT and HDS greases. In order to understand the underlying mechanisms of this effect one must consider the structure and composition of clay thickeners.

Bentonite, as previously described in section 2.1.3.1 of this work, is a claybased thickener, which underlies a polarity-changing process of activation during the production. According to [41], the hydrocarbon alkyl-chain rests applied in this process highly influence the bonding character and the consequent thickening behavior in BT greases. For this reason highly polar oils more naturally attach to these hydrocarbon molecule chain rests in organosilicon particles of BT greases than low polarity oils. Thus, they interfere with and partially prevent the formation of the BT matrix. For this reason the use of the same BT thickener type, containing equal ammonium ions with similar hydrocarbon alkyl-chains, must result in individual thickener percentages for all base oils.

As can be seen in table 4.2, the formulation of HDS greases is also subjected to polarity influences. It is observed that highly polar oils show a higher tendency to dock with Si−OH groups, protruding from primary particles of HDS greases, than low-polarity oils. This causes a disturbance of the formation of the three-dimensional network and generally results in higher percentages of thickener needed to thicken highly polar oils with equal mechanical stability.

So far, two different polarity-related effects on greases and their characteristics have emerged as illustrated in figure 5.24. In this chart, curved filled arrows represent all direct influences, whilst slim straight arrows represent possible mutilations of direct influences. On the one hand, there is a direct influence on the base oil/thickener ratio even in the formulation of the greases, as discussed earlier. On the other hand, the structural breakdown of greases under transient shear is also directly affected by base oil polarities, as has been shown by means of the energetic evaluation of the transient shear results. This double effect makes it exceedingly difficult to reveal the exclusive influence of the base oil polarity on mechanical structural breakdown because this degradation is also essentially influenced by the thickener concentration. In order to disclose the exclusive influence of the base oil polarity on mechanical structural degradation it is necessary to establish measurement conditions, which solely reveal the same by keeping constant all other influencing factors. The bond forces evoked and influenced by base oil polarities are mainly Keesom forces, as discussed in section 2.2.2.1. They are the strongest secondary valence bond forces, hence, in the context of adjustment of measuring conditions, it seems plausible to find conditions, which

5. Experimental analysis

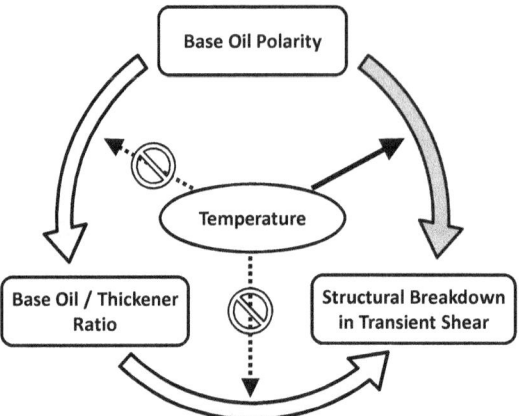

Figure 5.24.: Chart of different polarity influences on structural breakdown in transient shear—Note: Bold curved arrows represent direct influences, slim straight arrows (dotted and solid) represent possible mutilations of direct influences.

solely mutilate or even eliminate Keesom forces. It is well known that the formation of Keesom forces is highly temperature dependent [51, 64]. Through atomic scale oscillation evoked by a temperature increase, they are intensively reduced. In other words, the higher the medium temperature during the test, the lower are the Keesom forces in the grease microstructure. At low temperature conditions, however, Keesom forces are maximized. For this reason, the above-described temperature range from -10 °C up to 80 °C was chosen and polarity influences were isolated from other effects, such as percentage of thickener content. This is illustrated by the slim arrows in figure 5.24, pointing to the bold filled arrows representing a mutilation of these direct influences. This way, figure 5.24 makes clear that the temperature only disturbs the direct influence of base oil polarity on the structural breakdown. In reality, the temperature could also marginally mutilate the influence of base oil polarities on the thickener concentration. But this influence, as inherently limited, is only exerted in the moment of grease production and loses its effect after completion of the production. Moreover the temperature does not mutilate the influence of the base oil/thickener ratio on the structural breakdown in transient shear.

For further discussion of the investigation of temperature influences, all energy densities analytically determined by integration according to equation 2.39 are plotted as a function of temperature in the complete region of all measurement points as depicted in figure 5.25. In this plot, the grease systems with highly polar oils, TMPO and HOSO, are color-coded in black solid lines with different line tags for the respective groups of thickeners and base oils. All curves of greases based on low-polarity oils are color-coded with gray solid lines without any information on

5.4. Rheological analysis

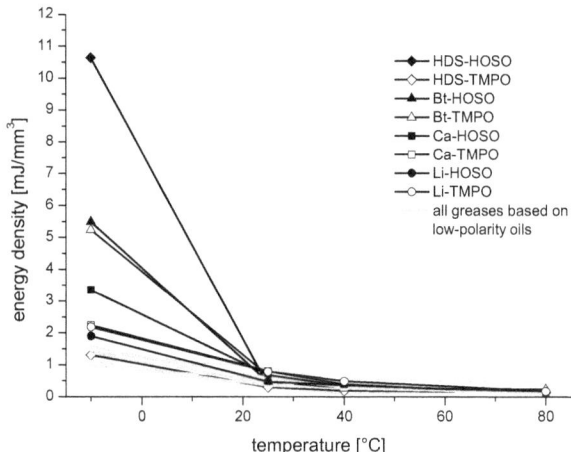

Figure 5.25.: Plot of energy densities $e_{rheo-rot}$ determined at all given temperatures

the thickener or base oil. This figure clearly shows that energy densities determined at temperatures higher than 25 °C all quite much resemble. The largest deviation from a linear trend mainly occurs in measurements performed at -10 °C. This deviation from a linear trend highly depends on the base oil polarity. While all energy densities determined for greases based on low-polarity oils stay within a narrow fluctuation band even at -10 °C, the greases based on highly polar oils do not show this behavior. In fact they even result in curves that are best fitted by an exponential approximation. The largest deviations from a linear trend are found for the grease systems formulated with clay thickeners. As expected, HDS-TMPO is an exception to this rule for the prior-discussed reasons. The greases formulated with metal soaps and highly polar oils measured at -10 °C only slightly but still clearly result in higher energy densities than all other greases based on low polarity oils.

This apparent exponential temperature dependent behavior of energy densities leads to the assumption that it may follow an Arrhenius-type fitting, which generally describes the temperature dependent kinetic constant in chemical processes. Adaption of the chemical Arrhenius-type fitting to the here-investigated physical process of temperature dependent degradation under transient shear leads to

$$e_{rheo-rot} = A \cdot e^{\frac{E_a}{R \cdot T}} \qquad (5.1)$$

where T $[K]$ represents the temperature, R represents the universal gas constant $(8.314 J K^{-1} mol^{-1})$, A is a dimensionless pre-exponential factor and E_a $[J \cdot mol^{-1}]$ is an activation energy, which in this case, has the physical meaning of the input energy necessary to produce structural degradation in the grease sample under transient shear. Plotting the results of Arrhenius fits in logarithmic scale over the reciprocal temperature reveals the ratio of E_a and R as the slope of the function. Figure 5.26 depicts the original data along with the corresponding Arrhenius-type fits for some

5. Experimental analysis

selected highly polar and non-polar greases.

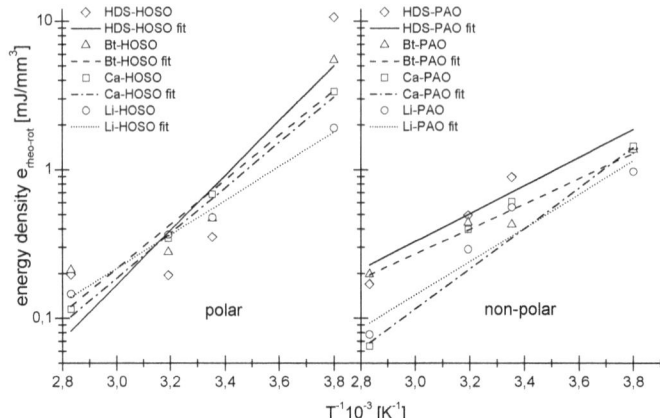

Figure 5.26.: Original data and Arrhenius fits of greases based on highly polar HOSO and non-polar PAO

This figure clearly depicts a constant quality of the adapted Arrhenius fits, which is found in the good correlation of the linear Arrhenius graphs and the scattered dots of the respective original data. Only HDS-HOSO shows a significant deviation between the fit and the original data. This figure, directly confronting both polarity groups, also shows a clear distinction between greases formulated with highly polar and low-polarity oils. It appears that the highly polar oil-based grease systems require a larger portion of activation energy in order to produce structural degradation in the grease samples under transient shear. This presumption is confirmed as its effect is depicted very well in figure 5.27 where all values of activation energy E_a were isolated from the analytically determined fit function. All values of E_a are displayed in groups of thickeners. In all of these groups a polarity influence more or less severely affects the activation energy, resulting in higher activation energies necessary to evoke structural degradation in greases based on highly polar oils than in those based on low polarity oils.

This effect most significantly results in larger differences between greases based on high and low-polarity oils within the groups of clay thickened greases HDS (38.5 %) and BT (52.7 %). HDS-TMPO, for explained reasons, is presumed to have resulted in much higher activation energy than displayed by the found data. A polarity influence is also found in both soap thickened grease systems although its effect results in smaller differences between the activation energies of high and low polarity oils-based greases—Ca (13.0 %) and Li (5.9 %). This effect may be explained by the stronger bond forces evoked between the clay thickener matrix and the highly polar oils surrounding it. It is presumed that the clay thickeners, which all required more activation energy to commence the structural breakdown in combination with highly polar oils, more naturally interact with the same. Low-polarity base oils consequently result in less activation energy needed in the process of structural breakdown. It is

5.4. Rheological analysis

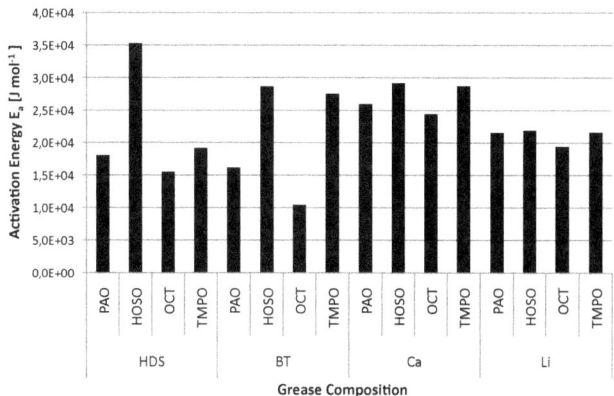

Figure 5.27.: Activation energy E_a, in transient shear flow

generally noted that clay thickeners much more sensitively react to a change of oil polarity in lubricating greases. Reasons for this effect are found in the microstructure and the physical mode of action of all investigated thickeners. In order to elucidate this behavior, one needs to understand the physical functionality of the investigated thickener systems at the molecular as well as the macro level. This especially applies if considering that only the synergy and interaction of the different grease ingredients on the mentioned levels cause the overall transient flow response. This micro structural analysis has successfully been performed in section 5.2. There the fibrillar soap structure of Li greases with single, partially interwoven and entangled loops of different size as opposed to the densely and multi directionally arranged solid particles of BT greases and the very small size of HDS primary particles, which form a rather rigid system have been shown and discussed.

In addition, the AFM analysis performed in section 5.2 allows to elucidate the physical mechanism acting in the above-discussed transient shear flow as well as the mixing ratio of the greases. Hence, a lubricating grease based on a chemically bound metallic soap with all its fibrillar, interwoven and entangled loops is expected to respond differently to changing oil polarities than a clay-grease with its micro platelet structure. This occurs although both thickening systems only physically interact with the surrounding base oil and with each other.

The mentioned AFM analysis in section 5.2 also showed that the base oil polarities in combination with Li thickeners do exert an influence on the thickener/base oil ratio. This had been shown especially with AFM images of soap greases depicted in figures 5.1, 5.2 and 5.3. Longer and less entangled fibers with less loops per examined unit of volume had resulted from the combination with highly polar oils TMPO and HOSO as compared to the low polarity base oil OCT.

In addition, the fact that base oil polarities only marginally affect the activation energy needed for structural breakdown in transient shear, which might also be labeled a thickener-share-adjusted degradation behavior, leads to the presumption

5. Experimental analysis

that they exert their influence on Li greases mainly once—during the formulation of the grease. As soon as the Li soap is completely formulated the indirect influence of base oil polarities on the transient flow response depends on the length, the type, the rate of entanglement and the ratio of soap fibers.

For technical reasons it was not possible to obtain accurate AFM micrographs of the examined Ca greases as discussed in 5.2. Hence, the response of their microstructure to alternating polarities cannot fully be elucidated. Because of the above-discussed polarity-independent constant thickener ratio (see table 4.2) it is presumed, however, that the structure of Ca greases is independent of oil polarity to a large degree. In combination with the relatively high polarity difference in the thickener-share-adjusted evaluation of grease degradation by means of activation energy it is presumed that the structural breakdown of Ca-thickened greases is mainly subject to polarity influences.

As previously mentioned, the platelets of the clay particles are considerably smaller than the fibers of the soap greases. They can only physically interact with each other in the formation of the supporting matrix structure. The smaller the particles, the more bonding sites are available per considered volume unit. Therefore, it is presumed that they evoke greater polarity-dependent rheological shear stress responses in the grease. Furthermore, this is substantiated by the fact that the interaction of the solid body particles only occurs through the surrounding base oil in the formation of the support network. In other words, the base oil works as adhesive between the particles. This is also the case for soap fibers, yet they are significantly longer and thus, although naturally offering a greater target surface per fiber in this adhesion process, they offer less target surface per considered unit of grease volume. In addition, soap fibers are much more flexible than clay particles, and thus they offer a greater contact surface even in sliding produced by mechanical shearing. In summary, it is presumed that the rheologically measured polarity-dependence of clay greases is evoked by the small size of clay particles, their solid body characteristics and its associated lack of flexibility. This would also elucidate why, with the activation energy, the clay-thickened grease systems, HDS and BT, showed greater polarity dependence than the Li and Ca soap-thickened greases.

5.4.1.2. Intermediate conclusions on rotational transient tests

An attempt has been made to disclose the exclusive polarity influence on the structural degradation in transient shear by means of Arrhenius-type fitting of energy densities. This approach constitutes a thickener-share-adjusted consideration of polarity influences. In comparison of this approach to the sole evaluation of energy densities in transient shear, which is still vitiated by a thickener-share influence, one finds that the extend of influences resulting from the activation energy corresponds more to the energetic behavior in low temperature conditions (-10 °C). On the other hand, the pure thickener ratio is best reflected by the energetic evaluation at high temperatures (40 °C and higher). This is not surprising because the Keesom-forces, and consequently the influence of polarity, diminish with higher temperatures, which in turn results in a predomination of the influence of thickener ratio. The fact that the energetic evaluation at low temperatures more corresponds to the results found with the activation energy indicates that in it polarity influences clearly predominate.

The results of these tests demonstrate that there are really two different base oil polarity influences on the rheologically measured structural degradation of lubricating greases in transient shear flow. On the one hand, a polarity-dependent influence on the thickener/base oil ratio was detected. This, in turn, exerts a unique influence on some rheological characteristics. On the other hand, the base oil polarity, considered alone, does also influence the rheological grease behavior in transient shear. This was found with an Arrhenius-type approximation of the temperature-dependent function of energy densities in combination with the temperature-dependent formation of Keesom forces. Depending on the applied thickener, these tendencies show different intensities. In a thickener-share-adjusted interpretation of the energy density temperature dependence, by means of Arrhenius-type fitting, it is found that, in fact, both polarity-induced influences are exerted in the clay-thickened systems. Within the group of clay-thickened greases, the highly polar oils increase not only the percentage of thickener but also the activation energy necessary to produce structural degradation. This double polarity influence, although not as severely, is also found in the group of Li soap-thickened greases. Here also, the thickener ratio and the level of activation energy needed for structural breakdown are affected by base oil polarities. In the Ca greases, however, there is no detectable influence on the thickener/base oil ratio, whilst there is a slight tendency pointing to a polarity dependence of activation energy in the mechanical degradation.

5.4.2. Amplitude sweep tests

In the previous chapter it has been proven that base oil polarities exert a substantial influence on the rheological behavior of lubricating greases in transient shear flow. This was done by means of energetic evaluation of rotational transient measurements. It is suggested to substantiate the found effects with other rheological testing methods and to validate if they might lead to comparable results. This proposal is now ensued by means of oscillatory shear tests.

5.4.2.1. Results and discussion of amplitude sweep tests

The data collected with the rheometer was exported and plotted in a diagram as can be seen in figure 5.28. This diagram shows the course of G' and G'' over $\gamma[\%]$ in duplex logarithmic scale. The unique appearance of each group of thickeners is displayed very clearly. The Ca-soap thickened greases resulted in a plot of the storage and loss moduli, which quite much resembles the one presented as standard in DIN 51810-2—they do not show a 'distinct hump' in the course of G'' right before the decline to the cross-over point. All other greases do show this 'distinct hump' especially the HDS-greases. At first glance, it is remarkable that the two groups of soap-thickened greases seem to be less susceptible to notice base oil influences. This becomes apparent by the relatively narrow band of G' and G'' values for these two greases. This effect is best observed in the group of Ca-greases, but also the Li-greases show it with the exception of Li-PAO. The clay thickened greases, especially the group of HDS-greases, exhibit a widely spread band of data plots of

5. Experimental analysis

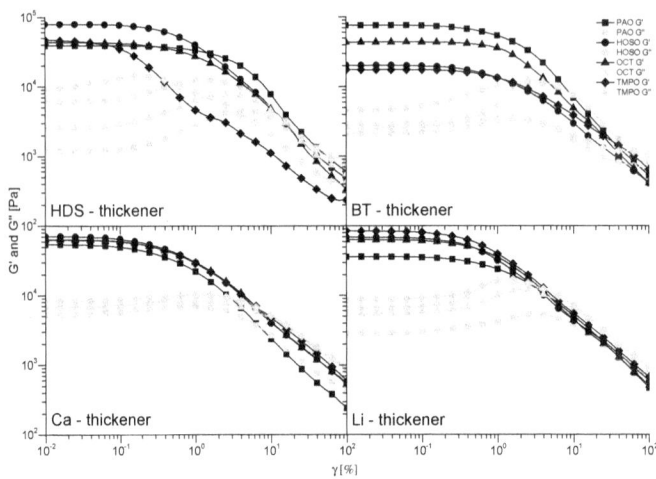

Figure 5.28.: Dataplot of G' and G'' of all 16 greases sorted by thickeners—amplitude sweep

G' and G''. The plateau values of G' in the LVE-range averaged for each group of thickener all amount to a similar value. The two soap-thickened greases, with values of 63.0 kPa (Li) and 61.5 kPa (Ca), are very close to each other and mark the upper region of all investigated thickener groups. Slight differences are found for the clay greases with 51.9 kPa (HDS) and 39.7 kPa (BT) in the lower region. The development of values of the loss modulus G'' behaves quite similarly. The fact that greases with high values of G' also result in high G'' values can be observed as a general tendency. In comparison of these two it should be mentioned that values of G'' are much smaller than those of G'. Values of G'' in the LVE-range are located at 5.84 kPa (Li), 6.69 kPa (Ca), 4.76 kPa (HDS) and 3.01 kPa (BT). The fact that the difference between G' and G'' is of the order of magnitude of one decade proves that the elastic behavior prevails over plastic-viscose behavior within the range of linear viscoelasticity.

The LVE-plateau G' values of the single greases not averaged in combined groups of thickeners help to reveal base oil influences. For a better comparability the G' values in kPa determined in the LVE-range are displayed in a bar diagram sorted in groups of thickeners in figure 5.29. This figure very clearly exhibits the base oil influences. The insights relating to polarity influences of the base oils on mechanical grease stability gained in section 5.4.1 invite to interpret the present data in a like manner. Here too, the unique polarity influences appear in tendencies as can be seen in figure 5.29. It shows that highly polar oils, HOSO and TMPO, tend to result in higher values of G' than low polarity oils PAO and OCT. This influence is not as distinctly depicted as it was with rotational transient tests. The results of the determined values of G' in the group of HDS greases, however, quite much resemble the rotational results. Here too, the influence of HDS-TMPO is not exerted as

5.4. Rheological analysis

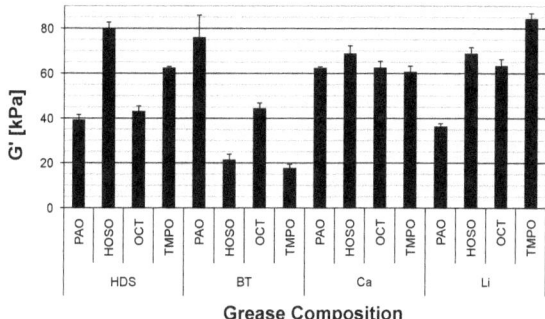

Figure 5.29.: Diagram of the storage modulus G' of all 16 greases sorted in groups of thickeners—values determined in the LVE-range

strongly as the one of the other highly polar oil HOSO. Also the storage moduli G' determined in the group of Ca-thickened greases insinuate this effect with Ca-TMPO deviating from its predicted tendency. The group of Li-greases, on the other hand, shows the effect of this influence quite clearly with all greases based on highly polar oils resulting in larger values of G' than those based on low-polarity oils. The only exception to this rule of linking high base oil polarities to high values of G' is the group of BT-thickened greases. Here this influence appears completely reversed with low-polarity oils PAO and OCT resulting in the highest values of G' compared to the highly polar oils HOSO and TMPO.

A comparison of the values of the loss moduli, G'', of the single greases may help to discover base oil influences just as it did for G'. The mentioned results of G'' values determined in the LVE-range are displayed in the bar diagram of figure 5.30. Results are again sorted by groups of thickeners and displayed in kPa. This diagram

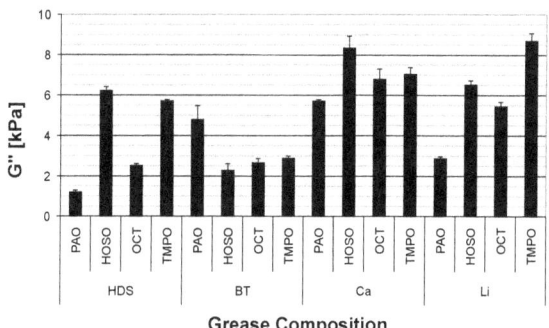

Figure 5.30.: Diagram of the loss modulus G'' of all 16 greases sorted in groups of thickeners—values determined in the LVE-range

5. Experimental analysis

reveals that base oil polarity influences are exerted more intensely on the loss moduli G'' than on the storage moduli G'. In the group of HDS-greases, the G'' results determined for greases based on highly polar oils, HOSO and TMPO, very distinctly segregate from those formulated with low polarity oils PAO and OCT. The values of G'' in the group of Ca- thickened greases exhibit more than just the tendency to show the same polarity-dependent behavior that was detected for values of G'. The G'' value of Ca-TMPO is, not extraordinarily, but still higher than those of the Ca-greases based on low-polarity oils. The group of Li-greases, too, exhibits clear polarity influences on the loss moduli G''. Hence, Li-HOSO and Li-TMPO result in significantly higher values of G'' than Li-PAO and Li-OCT. The very distinctive size difference of G' values based on polarity influences in the group of BT-greases has diminished along with the reversed polarity effects in the evaluation of the loss moduli G''.

A comparison of the polarity-induced influences on the oscillatory response of greases illustrates that base oil polarities exert a more substantial influence on the loss moduli G'' than on the storage moduli G'. Therefore, it is deduced that base oil polarities seem to have a greater influence on the viscous than on the elastic behavior of lubricating greases. This would explain why polarity influences are more distinctively detected in rotational than in oscillatory rheological tests, since this type of rotational tests, if neglecting the initializing effects investigated by Delgado et al. [30], only examine the transient viscous behavior.

It is one thing, to separately evaluate the magnitude of the storage and loss moduli G' and G'', respectively but additionally it appears interesting to evaluate the relative elasticity of the greases or the tangent of the loss angle δ. Figure 5.31 shows values of $tan(\delta)$ of all greases determined in the LVE range. As can be

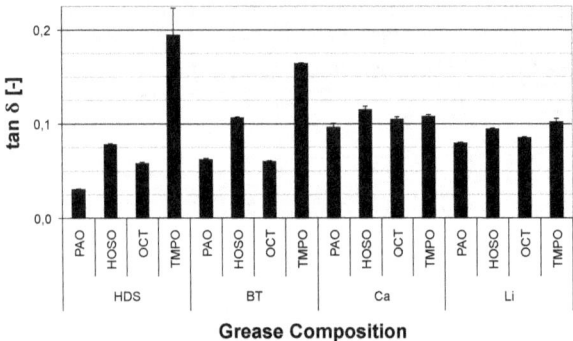

Figure 5.31.: Loss tangent, $tan(\delta)$, of all 16 greases sorted in groups of thickeners—values determined in the LVE range

observed in this figure, all values of $tan(\delta)$ are located clear beneath 1—a fact that could have been expected in the LVE range. What also appears very interesting in this figure it the fact that within all groups of soap-thickened greases all values of the loss angle are very stable around $tan(\delta) = 0.1$, whereas both groups of clay-thickened greases resulted values of $tan(\delta)$ ranging from 0.03 up to 0.19. Additionally, base oil influences in all thickener groups are detected very clearly in figure 5.31 leading to higher values of $tan(\delta)$ in all greases based on highly polar oils TMPO and HOSO than in low polarity oils PAO and OCT within each respective group of thickeners. In other words, the higher the base oil polarity the stronger is the propensity toward a plastic deformational character in amplitude sweep tests. This effect appears very clearly with all formulated thickeners but its influence is much stronger in the two groups of clay-thickened greases resulting in differences of $tan(\delta)$ of 0.18 (HDS), 0.15 (BT), 0.02 (Ca) and 0.03 (Li). The differentiation between base oil polarities of each grease system appears more clearly in the evaluation of the loss tangent than in the evaluation of G' and G''. Additionally, this figure shows very clearly that the evaluation of $tan(\delta)$ does not result in a reversal of the polarity influence compared with rotational transient results in combination with BT-thickened greases as did the evaluation of G'. This circumstance could also have been derived from the fact that G'' values of all BT greases present themselves very stable in the in LVE range whilst G' showed a relatively strong dependence on base oil polarity.

According to section 2.5.2, a suitable method needs to be selected to determine the limits of the LVE-range of the single groups for comparison of the different deformation dependent changes of grease consistency. The evolution of G' and G'' were each analyzed to detect a change larger than 10 % in order to determine the limits of LVE-ranges in the present data. Therefore, the quantity first deviating by more than 10 % marks the end of the LVE-range. In the evaluation of the present data, the end of the LVE-range in most cases was defined by a first decline of the storage modulus G'. However, the HDS-greases, which in all cases exhibited the above-mentioned 'distinct hump' in the course of G'', make an exception. In this group, the end of the LVE-range of all greases is detected by an incline of the loss modulus G''. Except from the HDS-greases there are only two more in all of the sixteen greases, which exhibit this behavior. These are BT-PAO and Li-HOSO. A comparison of the deformation values defining the end of the LVE-region of all greases averaged in groups of thickeners results in 0.04 % (HDS), 0.29 % (BT), 0.1 % (Ca) and 0.09 % (Li). These values clearly show that the LVE-range of BT-greases is wider compared to all other greases. Since the deformations of the single greases do not deviate significantly within their respective group of thickeners they will not be discussed any further.

Analogous to the evaluation of the LVE-range extension, the evaluation of the cross-over points outside the LVE-range, will lead to information about base oil influences too. At the cross-over point the value of the loss modulus equals the value of the storage modulus. The diagram displayed in figure 5.32 shows all these values grouped in types of thickeners.

The displayed values of $G' = G''$ in the cross over points determined for the given greases also exhibit the prior-discussed influence of base oil polarities on the rheological behavior. This results in distinctively higher values of both, strain and stress moduli of all HDS-, Ca- and Li-thickened greases based on highly polar oils, HOSO and TMPO, compared to the greases based on low-polarity oils PAO and OCT. Once again, the group of BT-thickened greases makes an exception to this rule by perfectly reversing the found polarity effect to the opposite. In addition,

5. Experimental analysis

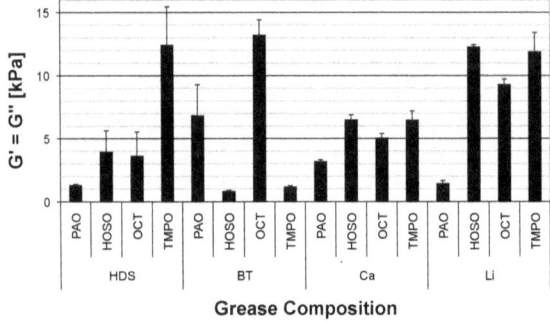

Figure 5.32.: Values of $G' = G''$ of all 16 greases sorted in groups of thickeners—values determined in the cross over point

this effect causes significantly large differences between values determined in the cross over points for highly polar greases (BT-HOSO = 0.77 kPa and BT-TMPO = 1.17 kPa) and those determined for low polarity ones (BT-PAO = 8.11 kPa and BT-OCT = 10.87 kPa). With these results the measured $G' = G''$ for BT-greases based on low polarity oils amounts to 10.22 % of $G' = G''$ value in the greases based on high polarity oils.

In addition to the values of G' and G'' also the corresponding deformations in the crossover points can be evaluated. A depiction of these deformation values in % is found in figure 5.33 where the familiar and frequently used classification into groups of thickeners helps to compare the results to those of other evaluations. This

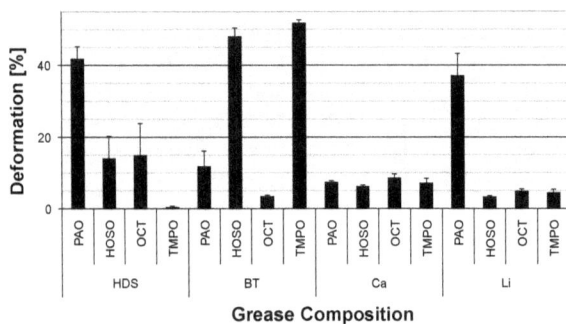

Figure 5.33.: Deformation in % of all 16 greases sorted in groups of thickeners—values determined in the cross over point

figure also shows an explicit base oil influence on the deformational behavior of the investigated lubricating greases. It is noticeable that greases, which exhibited large shear moduli at the crossover point result in small deformation in the same point

5.4. Rheological analysis

and vice versa. This reversal does not come into effect with some selected greases only but it pertains to all greases investigated in the course of this experiment. It is even performed in such a perfect manner that all effects found with the shear moduli are completely reversed to the opposite in the deformation. To a certain degree, this effect can be elucidated by the general course of the shear moduli. Greases, which in oscillatory shear with small amplitudes appear relatively rigid and exhibit high moduli of shear tend to yield to shear stress sooner than comparable softer greases with small shear moduli. The found results point to the inadequacy of the consistency measurements by cone penetration according to DIN 51818 and ISO 2137. Although shear tensors in cone penetration tests substantially differ from those prevailing in amplitude sweep tests, because they are not measured oscillatorily, in this kind of consistency test the grease is deformed elastically and plastically. This results in shear sequences in cone penetration tests, which are simulated by amplitude sweep tests in the LVE-range, the crossover point and beyond. Because all greases were formulated with the same NLGI grade, the oscillatory results in the LVE range need to be superimposed on the results obtained in the crossover points in order to give the same information. This means that if, in inversion of this argument, greases of the same consistency grade are examined by oscillatory measurements the deformational results harmonize the results of different shear moduli.

The above-given figures 5.28 to 5.31 and the accompanying disquisition do not give any information about the energetic comparability of these present results to the ones obtained with rotational rheological tests. For this purpose, the energy densities of the distinctive points in the curves (LVE-range and crossover point) need to be determined. Following equation 2.50 the energetic evaluation of amplitude sweep tests requires the physical quantities $G'[Pa]$, $\gamma[m/m]$, and $cos(\delta)[-]$. At first the deformations at the distinctive points were determined with the rheometer's controller software. These were used to calculate the respective data, which was subsequently exported into a spreadsheet in order to determine the corresponding values of energy densities. Figure 5.34 displays the calculated energy densities e_{rheo} of the single greases obtained at strain values of $0,01$ in the LVE-range in J/m^3. The hight of the bars in this diagram quite exactly resembles the hight of the bars representing the storage moduli displayed in figure 5.29. This is obvious even though different physical quantities apply in both images. For one thing, this similarity is based on the fact that in the beginning of the LVE-range all deformations amount to a similar value close to zero. For another thing all values determined for $cos(\delta)$ approximately amount to 1 because of the maximum difference between G' and G'' right in the beginning of the LVE-range. Insertion of this knowledge into equation 2.50 delivers

$$e_{rheo} = \frac{G' \cdot \sim 0,01^2}{\sim 1} \qquad (5.2)$$

and most likely results in similarly shaped bars that are reduced by a common factor. The displayed results reveal an influence of the base oils and the thickeners on the rheological behavior. But since the tendencies are equal to those disclosed in the evaluation of the storage moduli G' in the beginning of the LVE-range only reference to the same shall be given here. In application of the there given discussion on

5. Experimental analysis

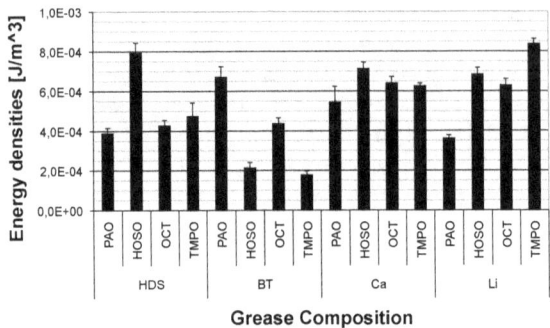

Figure 5.34.: Energy densities of amplitude sweep tests—LVE-range

polarity-induced influences between base oils and rheological behavior it can only be repeated that HDS-, Ca- and Li-greases based on highly polar oils result in higher energy densities compared to the same greases based on low polarity oils. Here also this effect is reversed to the exact opposite within the group of BT-greases.

The energetic evaluation of amplitude sweep tests in the crossover point can be performed just like the evaluation in the beginning of the LVE-range. Again the energy densities of all greases were determined in a spreadsheet with the prior exported values of $G' = G''$, γ and $cos(\delta)$ and are displayed in the known manner in groups of thickeners in figure 5.35. As γ describes the deformation of the completely changed sample structure, the same data could have been derived using equation 2.51.

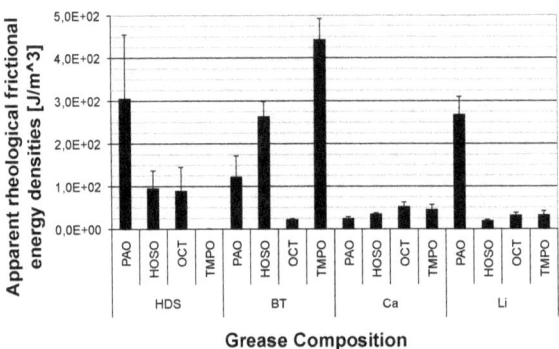

Figure 5.35.: Apparent rheological frictional energy densities of amplitude sweep tests—Crossover point

In this figure attention is drawn to the clearly visible base oil influences on rheologically measured energy density even in the crossover point. Compared to

5.4. Rheological analysis

other tests, it is striking that the results displayed in this figure not only tendentiously but exactly resemble those of the deformation in the crossover points. This effect may be elucidated by equation 2.50, which was used for the determination of the rheological energy density. The values of G' inserted into this formula still exhibit the above-mentioned well-known polarity influences on the storage modulus in the crossover point. Because of $G' = G''$ values of $cos(\delta)$ will naturally result in $cos(45°)$. But the relatively high values of deformation in the crossover points exert their quadratic influence in the formula toward a tendency to more likely describe the deformation than the shear moduli. Since these results show so clearly that energy densities in the crossover points correspond to the deformations in the same points one can also relate the base oil polarity influences on rheological behavior found in deformations back to energy densities. This is shown in the HDS-, Ca- and Li-greases based on highly polar oils, HOSO and TMPO, resulting in smaller energy densities than greases based on low polarity oils. This effect is reversed again in the BT-greases.

5.4.2.2. Intermediate conclusions on amplitude sweep tests

In summary, the following effects were revealed from amplitude sweep tests. First, it was shown that soap thickened greases, averaged in their respective group of thickeners, generally resulted in higher values of G'. This may be attributed to the fibrous soap structure. Its long interwoven fibers are presumed to evoke a higher elastic deformability. The clay-thickened greases in comparison are not interwoven and consequently they slide off more easily. Polarity effects, however, may increase the elastic deformability of both thickener types.

Another finding of the amplitude sweep tests are similar polarity influences with HDS-, Ca- and Li-greases in both the beginning of the LVE-range and the crossover point as were found in rotational tests. Only the polarity influences of BT-greases are opposite to those found with rotatory tests. These insights are exhibited more clearly in G'' values than in G' values except for the reversal in BT-greases.

It was also found that the LVE-range of most greases is limited by a decline of G'. Especially the HDS greases, which all showed a 'distinct hump', made an exception by defining the limit of LVE-range by an incline of G''.

Another finding of the performed amplitude sweep tests is the fact that deformation values in the crossover point resemble the reciprocal of the values of $G' = G''$ at the same point.

It was also found that, on the one hand, the evaluated energy densities in the beginning of the LVE-range reflect the behavior of G' values including all polarity influences that were found. On the other hand, the energy densities in the crossover points reflect the behavior of the deformation.

The most interesting conspicuousness in the results of the performed amplitude sweep tests, is the reversal of the found polarity effects in the BT-greases. In almost all evaluations of amplitude sweep tests it was noticed that base oil polarity exerts a large influence on the rheological behavior of lubricating greases. These influences work in a way that, just as in rotational tests, greases based on highly polar oils evoke larger energy densities. In the same way, it was discovered with amplitude sweep tests that these polarity influences appear reversed in the group of BT-greases. There are several findings that help explain this effect.

First of all, the available results of amplitude sweep tests give reason to verify if the thickener structure in combination with base oil polarities may elucidate this effect. This seems obvious because the applied thickener structures fundamentally differ from each other. In the explanation of this effect the microstructure of the different thickeners need to be realized. These were already discussed thoroughly and graphically displayed with AFM micrographs in section 5.2. The figures displayed there, as they highlight their differences, are used again to elucidate the mutual interrelation of the thickeners' microstructure and rheological effects. As already explained in section 5.2, the Li-soap thickener is structured with single, partially interwoven and entangled loops of different size. BT-greases, on the other hand, are structured with relatively large, densely and multi directionally arranged platelet particles. The structure of HDS-greases consists of a framework of very small cross-linked circular primary particles. Even though it was not possible to take AFM micrographs of the Ca-soap structure it is still presumed that these, just like Li-soaps, consist of a fibrous structure. Farrington [38] describes these and reports of Ca-soap fibers in grease thickeners that are shorter than Li-soap fibers. If greases with the mentioned different microstructures are submitted to oscillatory stresses, different rheological responses have to be expected. In oscillatory amplitude sweeps the amplitude-dependent rigidity of the greases is to be taken into consideration. The LVE-range, in which elastic deformation prevails, might evoke different effects compared to the crossover point and its maximum plastic deformation.

Secondly, a direct comparison of the energetic evaluation of amplitude sweep tests to the same evaluation in rotational transient tests is only possible to some extend. This is valid because of the different directions of movement. Oscillating amplitude tests evoke alternating stresses within the grease structure, whereas in rotational tests a constant shear rate is applied, reflected in a transient continuous stress during the whole course of the test. These different deformation situations will inevitably evoke different reactions of the greases than in rotational tests.

As a third reason for the reversal of the polarity effects in combination with BT-greases it is presumed that the percentage of thickener plays a very important role in this regard. The evaluation of the energy densities of rotatory tests was adjusted from the thickener share. This was not done in the evaluation of the amplitude sweep tests. In the latter it was found that the polarity effect is mainly reversed in the magnitude of G' values, whereas the G'' values all appeared stable in the group of BT greases within the LVE range. On the one hand, this fact influences the evaluation of the energy densities from a purely mathematical point of view. Equation 2.50, which was used for the calculation of energy densities in amplitude sweep tests, only considers G' and leaves G'' completely disregarded. Therefore, if values of G' underlie the mentioned reversal of polarity effects then this will also be reflected in the calculation of the energy densities. On the other hand, it is presumed that the thickener percentage more severely influences the elastic behavior of the BT greases. This has been proven by the evaluation of the loss tangent, $tan\delta$. There, it has been shown that the reversal of the polarity effect in BT-greases was completely negated, resulting in higher loss tangents of all greases based on highly polar oils within their respective groups of thickeners. This fact justifies the conclusion to consider the loss tangent as a quasi thickener-share-adjusting magnitude, but only in combination with BT-greases.

5.4.3. Frequency sweep tests

In the previous chapters it has been expounded that base oil polarities exert influences on the rheological behavior of lubricating greases. This was done with rotational measurements and oscillatory test in amplitude sweep. Now an attempt will be made to deepen the knowledge gained by applying another oscillatory measurement method. For this purpose oscillatory rheological tests with constant amplitude and inclining frequency, so called frequency sweeps, will be performed. For the selection of the preset amplitude it is resorted to the data gained in the previous section.

5.4.3.1. Results and discussion of frequency sweep tests

The results of the performed frequency sweep measurements are depicted in figure 5.36 in terms of G' and G'' vs. frequency plots in log-log scales. Since the data presented

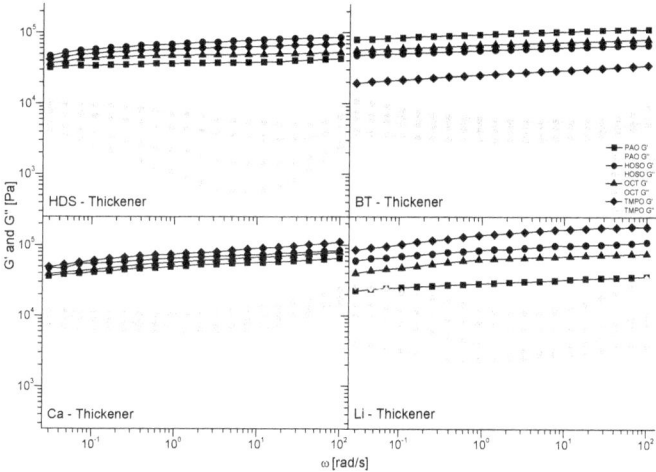

Figure 5.36.: Frequency sweep dataplot of G' and G'' of all 16 greases sorted by groups of thickeners

in this figure only represents an excerpt of the G' and G'' curves in amplitude sweeps it is not surprising that it reflects a similar behavior. These mechanical spectra inside the LVE range with values of G' around one decade higher than G'' and a minimum in the latter at intermediate frequencies reveal the aforementioned typical characteristics of polymeric systems with physical entanglements and give evidence to the highly structured systems that have been priorly revealed by means of AFM investigation. Additionally, these graphs present themselves with a certain degree of similarity to each other and to commercially available grease systems.

5. Experimental analysis

As can be observed in figure 5.36, the course of the loss and storage moduli of all greases depends on the applied thickener type with more or less similar band widths in all groups of thickeners as has been observed with amplitude sweeps. This is not surprising as the structure of the grease must inevitably depend on the thickener and, additionally, all SAOS frequency sweeps were performed in the LVE range.

As G' values do not change a lot with frequency and as the courses of G'' over frequency depict a somewhat symmetric behavior, it is useful to average the course of the plots for the whole region of applied frequencies for reasons of comparison of the data of the single greases and to present the results in a bar diagram similar to those already used in the evaluation of amplitude sweeps. This is done in figure 5.37 where all values of G' and G'' are presented in this manner in logarithmic scale. This scale was used in order to be able to present G' and G'' all together in one diagram. This figure clearly shows that the tendencies of G' and G'' resemble each

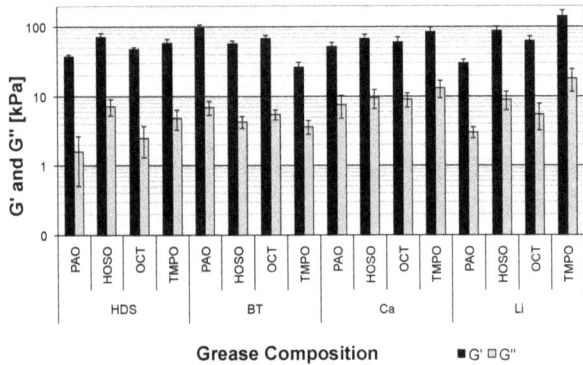

Figure 5.37.: G' and G'' determined with averaged values of frequency sweeps

other with quite accuracy. The only difference is that, as previously mentioned, G'' values are much lower, but other than that both values describe the same behavior for each group of thickeners. Again, this result is in no way surprising, since it only reflects the outcome already discussed in amplitude sweeps. There too, in the LVE range the behavior of G' and G'' shown at 1 Hz, resembled quite exactly. Therefore, the dependence of the greases' rheological behavior on the formulated base oil is also detected with frequency sweeps. In fact, most thickener groups even show the same ranking that was found with amplitude sweeps. Consequently, this ranking reveals the typical subdivision into the two known categories of highly polar oils HOSO/TMPO and the two low polarity oils PAO/OCT with its accompanying orientation in the respective thickener groups. This means that highly polar oils HOSO and TMPO as base oils in soap thickened and HDS greases evoke higher values of G' and G'' than the low polarity oils PAO and OCT. In the group of BT-greases this effect is reversed to the opposite as it happened in amplitude sweeps. Again, the similarities between the results of amplitude and frequency sweeps do not appear unexpected for aforementioned reasons.

5.4. Rheological analysis

The energetic evaluation of the determined data was performed in accordance with section 2.5.3 in the same way as it was performed in amplitude sweeps. The calculatory basis for the determination of energy densities was derived from the averaged values of the storage modulus $G'[Pa]$. The values of strain $\gamma[m/m]$ were output by the controller software of the rheometer as a function of the preset stress values $\tau[Pa]$. It also determined the values of the cosine of loss angle, $cos(\delta)$. All thus determined values of energy density are displayed in figure 5.38 in linear scale. This figure very clearly shows the individual responses of energy densities of each

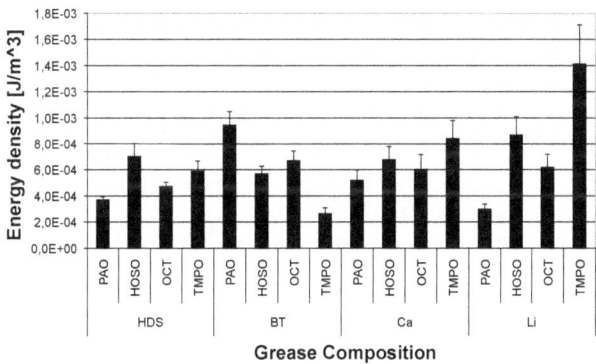

Figure 5.38.: Energy densities determined with frequency sweeps

single grease. Evaluation of the same from the thickeners point of view reveals that these results quite well correlate to the results derived from amplitude sweep tests. Not only the energy densities reach a similar level in both of these SAOS test methods but also do the individual groups of thickeners show a resemblant result. The groups of soap-thickened greases all exhibit the well known base oil influence with higher values of energy density for those greases based on highly polar oils. This difference is observed even more clearly than it had been found with amplitude sweep tests. The biggest difference of energy densities between highly polar and low-polarity oil-based greases is found in the group of Li-greases amounting to $0.682\,\text{mJ}/\text{m}^3$. Also the group of clay-based greases exhibits a similar behavior in both SAOS methods, which results in a ranking within the group of HDS greases of higher energy densities for greases based on highly polar oilsAdditionally, the group of BT-greases reveals a complete reversal of the found polarity effects with greases based on highly polar oils now resulting in the least energy densities. This same result, has also been found with amplitude sweeps.

5.4.3.2. Intermediate conclusions on frequency sweep tests

The fact that the energy density results of frequency sweeps so clearly resemble those of amplitude sweeps could have been expected even before the test as both of these respective evaluations are based on the same rheological characteristics and

5. Experimental analysis

physical quantities. Moreover, both methodologies were applied in the LVE range, in which the grease structure remains almost unaltered.

5.4.4. Tensile tests

5.4.4.1. Results and discussion of tensile tests

Tensile tests were perfumed in the above-described manner. The resulting curves of normal force over gap width were numerically smoothed and averaged for all repetitions of tests for each grease. All resulting maximum normal forces of these tests are displayed in figure 5.39 sorted in the known manner into single base oils and groups of thickeners. This figure clearly shows that the measured greases behave

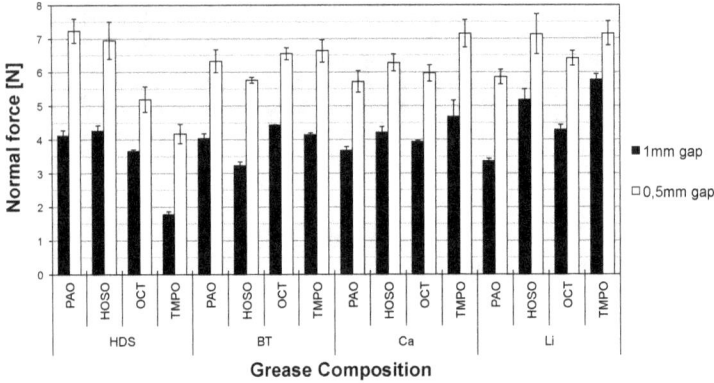

Figure 5.39.: Normal forces determined in tensile tests

quite consistently throughout all applied gap widths. A difference is found in the hight of the resulting normal forces of the measurements—the determined normal forces in gaps of 1 mm hight in average amount to 64.5 % of those measured with gaps of 0.5 mm. Another difference is found for the standard deviations of the determined values—in 0.5 mm gaps values deviate more than in the larger gaps. Evaluation of the present data from the point of view of the formulated thickeners reveals quite similar responses in all groups of thickeners. A ranking of averaged values throughout both measured gap widths results in HDS(4.67 N), BT(5.14 N), Ca(5.20 N) and Li(5.65 N) from the highest to the lowest. In this comparison it also becomes apparent that individual values in the groups of Li and HDS differ more from each other than in the groups of BT and Ca-greases. Analysis of the results from the point of view of the base oils reveals a consistent dependence on the applied base oils in all measured normal loads. The behavior in the groups of BT-, Ca- and Li-greases, in fact, quite much resembles the outcome of the rheological analysis in the beginning of the LVE range determined with amplitude sweeps. Here too, a distinct deviation into the known subcategories of highly polar oils HOSO and TMPO and low polarity oils PAO and OCT is generally found. In

5.4. Rheological analysis

the case of measured tensile stresses this difference is most distinctive within the group of Li-greases. Most interestingly, even the same reversal of polarity-induced behavior is found in the evaluation of gap widths in tensile test as it it was found in the mentioned amplitude sweeps. This results in higher values of normal force determined for highly polar oils HOSO and TMPO in the groups of soap thickened greases and in lower values of normal force within the group of BT-greases. These discussed similarities between tensile tests and amplitude sweep tests are not found in the group of HDS-greases. When considering the the group of HDS-greases on its own one finds a ranking that resembles the determined distinction into the subcategories of high- and low-polarity oils only in the measurements with a gap of 1 mm and only in the formulated base oils PAO, HOSO and OCT. The HDS TMPO grease, once again, makes an exception to the found rules.

The energetic evaluation of the so far discussed tensile tests was carried out with the formula given in equation 2.53. The integration of the normal force curve over elongation was executed in the rheometer's controller software. Values of zero for the start of the elongation and the individual maximum elongation of rupture were set as the boundaries in the mathematical integration of the normal force. All resulting values of energy density are displayed in the known manner in figure 5.40. This figure

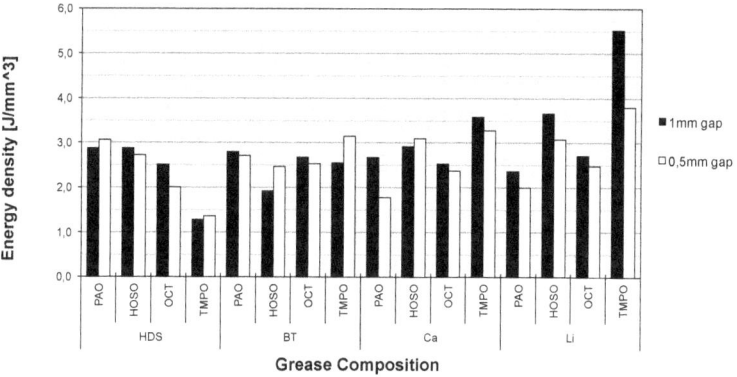

Figure 5.40.: Energy densities determined in tensile tests

depicts a very consistent behavior of all energy densities throughout all measured gaps in the tensile tests. In other words, individual responses of single greases found in the 1 mm gap are almost exactly repeated in the 0.5 mm gap. Analyzing the data presented in this figure from the thickeners' point of view reveals quite similar energy density responses in all groups of thickeners. In the values determined with both gap widths the group of Li-greases not only exhibits the largest deviations but also presents itself with the highest values of energy densities with an average of $3.56\,\text{J/mm}^3$. This ranking is followed by Ca ($2.93\,\text{J/mm}^3$), BT($2.48\,\text{J/mm}^3$) and HDS($2.39\,\text{J/mm}^3$). Analysis of the data presented in this figure from the base oils' point of view quite explicitly reveals the influences of the individual base oils in each group of thickeners. Generally, the base oil influences on the energy density

5. Experimental analysis

results are brought to light more distinctively in the tests performed with the smaller gap of 0.5 mm, but also with 1 mm gaps they are still detectable. The previously found subcategorization into groups of base oils with high and low polarity is also found in the energetic evaluation of tensile tests within the groups of soap thickened greases with this effect arising more clearly in the Li-greases than in Ca-greases. This influence is oriented in a way that the highly polar oils HOSO and TMPO generate higher energy densities than the low-polarity oils PAO and OCT. The BT-greases also reveal this effect but only in the tests performed with a gap of 1 mm and again with reversed orientations resulting in higher energy densities for the low-polarity oils PAO and OCT. Apart from that, this effect is not found in the group of HDS-greases in either investigated gap width. Here a clear ranking of energy densities is detected across all tested gaps that may be described by PAO, HOSO, OCT and TMPO from the highest to the lowest.

5.4.4.2. Intermediate conclusions of tensile tests

In almost all tests, this analysis unravels a distinct similitude with results of discussed normal forces. HDS-, Li- and Ca-thickened greases exhibit the same dependence on base oil influences. Only in BT-greases tested with the narrower preset gap of 0.5 mm the normal force analysis and the energy density analysis result in slight differences. This may be attributed to the impact of the different factors in the used formula. The volume of the greases inserted into equation 2.53, which was used to determine the energy densities, was at a constant level throughout all tests performed with equal gaps. Only the extend of normal force and the elongation-dependent behavior interpreted by means of integration influenced the results. Since the results of both analyses resemble as much as they do it is deduced that the courses of all curves of normal force over elongation also resemble.

A comparison of the discussed results to those achieved with other rheological tests such as rotational test, amplitude sweeps and frequency sweeps reveals some differences as well as similarities. The base oil influences in soap thickened greases, which resulted in high energy densities for all greases based on highly polar oils HOSO and TMPO, were detected with all rheological examinations regardless of the strain conditions. In case of tensile tests, this effect may be attributed to different stress tensor components compared to all other rheological tests and to the fact that the stress occurs absolutely instantaneous and only for a very limited time period. In long term rotational tests, as discussed in section 5.4.1 the investigated greases also undergo this instantaneous shear stress but it only occurs in the start of the test and results in the stress overshoot [30]. So in a way the results determined with tensile tests might be comparable to the initiating elastic deformation in rotational tests. The subsequent viscous deformation in rotational tests also takes place in tensile tests but it is not measurable as intensely since it only results a constricting process right before the rupture of the produced grease string. Since the grease structure is stressed so immediately in tensile tests it does not have any time to react to this stress by any flowing processes different from elongational flow and will in the moment of infinitesimal initiation mainly react with an elastic deformation of the inner network structure. This is the case especially with comparably narrow gaps. Therefore it is presumed that the higher differences between energy densities of high and low polarity oils in the 0.5 mm gaps measured for soap thickened greases result from a restriction of the flowing process evoked by narrow gaps. In case of soap thickened greases, the inner thickening network comprised of long and entangled fibers are presumed to be the part of the grease, which is elastically deformed before physical bonds break and the fibers slide off against each other. Since the network structure is stabilized by the surrounding base oil and its bonding features it is presumed that highly polar oils with their inherently stronger secondary valence bonds evoke higher energy densities in tensile tests. This theory may be corroborated by a comparison of the energy densities evoked by soap thickeners of different fibrillar lengths. For this reason it is presumed that Ca-greases result in less energy densities in tensile tests compared to greases thickened with Li-soap because of their shorter fibers. Just like in amplitude sweeps, tensile tests also result in a reversal of the polarity-induced effects in BT-greases. This effect in tensile tests, although contradictory at first sight, is presumed to be evoked by the stronger bonds of highly polar oils. Within the first fractions of a second of applied tensile stress the stronger bonds of highly polar oils are better able to align the BT-platelet structure parallel with the direction of enforced movement. Once these platelets are thus aligned they more easily slight off against each other as opposed to the platelets surrounded by low polarity oils, in which its weaker bonds were not able to align the platelets as presumed and modeled in the postulates made in section 3.

As opposed to all other rheological tests, with tensile tests no base oil polarity induced behavior was detected in the group of HDS greases. While this known polarity dependence might be interpreted into the tests performed with a gap of 1 mm only as a general tendency when disregarding the HDS-TMPO, there is no room for such an interpretation in the results determined with a gap of 0.5 mm. The absence of the known polarity dependent effects can only be induced by the HDS thickener structure. Therefore it is presumed that the HDS thickener structure with its very small spherical particles can not be aligned to the direction of motion in the first part of the induced movement. Moreover, owing to the small size of HDS particles the polarity-induced bonds work more multi directionally and thus they cancel each other in the instantaneous shear stress of tensile tests during the opening of the gap.

5.5. Other tests

5.5.1. Bouncing ball tests

Results of the analysis of filament length are depicted in figure 5.41. This figure clearly shows that each grease resulted in quite reproducible individual filament lengths. An analysis from the point of view of thickeners reveals that all deviations in the single groups quite much resemble with values ranging from 11.0 % in case of BT-greases up to 14.8 % in case of Ca-thickened greases. On the other hand, the grease filament lengths averaged in groups of thickeners vary largely with an increasing rank of HDS 7.79 mm, BT 8.30 mm, Ca 9.94 mm and Li 10.65 mm. Evaluation of the data presented in figure 5.41 from the base oils' point of view shows the accustomed

5. Experimental analysis

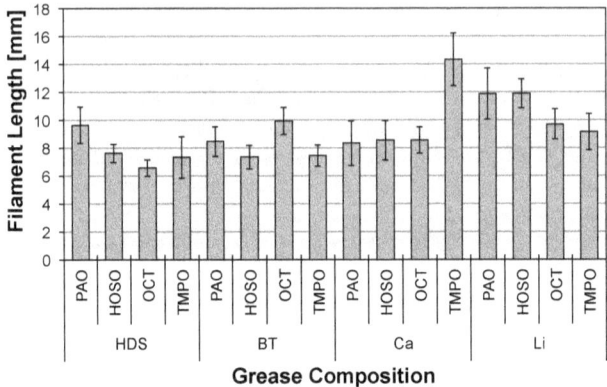

Figure 5.41.: Filament lengths measured in bouncing ball tests

classification into high and low polarity oils only in the group of BT-greases resulting in longer grease filaments for the category of greases based on low-polarity oils PAO and OCT. None of the other groups of greases based on the same thickener even remotely show this effect. Neither do they exhibit any behavior that could be interpreted into a common pattern. Concluding the evaluation of filament lengths created in bouncing ball tests, it should be mentioned that all greases showed very individual behavior not only in the resulting length of the filaments but also in other shape features of the filaments. Some greases, e.g. resulted in the creation of multi-filaments circularly arranged around the central contacting impact area with either two or more filaments. Other greases even resulted in a circular film instead of single filaments. Because of these highly different characteristics is seems problematic to relate the results to the micro structural setup of the greases.

The impact marks appearing as indentations in the evenly polished surfaces were also analyzed microscopically. On the one hand, only objectively measurable dimensions such as the mark diameter were determined. On the other hand, also solely subjectively determinable features of the impact marks such as the relative destruction of the surface were evaluated. Results of the analysis of imprint dimensions with its measured diameters are depicted in figure 5.42. This figure, too, shows a quite distinct individual and still reproducible behavior of all examined greases with very little deviation. The presented deviations only appear large because the value range of the y-axis starts with 500. In fact the deviations range from 2.5 % in case of BT greases to 3.9 % determined with Ca greases. Analysis from the point of view of thickener types reveals quite similar diameters for the groups of HDS (676 µm), BT(677 µm) and Li(681 µm). Only the Ca-greases differ a little more with an average of 705 µm. Just like with the evaluation of filament lengths, the diameters, too only reveal the well known dependence on base oil polarities in the group of BT-greases, when analyzing from the point of view of base oils. Also none of the other thickener groups show this effect.

What also becomes very clear at the examination of figure 5.42, in fact, is the

5.5. Other tests

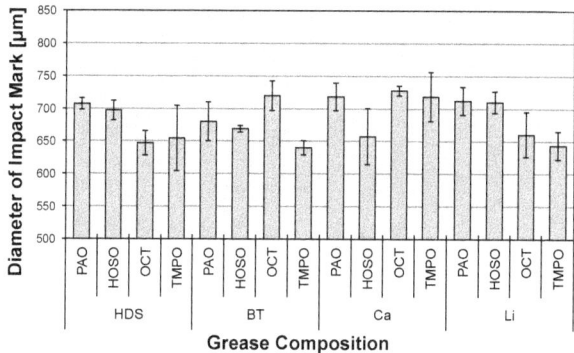

Figure 5.42.: Impact diameters measured in bouncing ball tests

good correlation between the diameters of the impact marks and the previously discussed lengths of grease filaments. For one part, this behavior may be described by the physical correlation between these two dimensions. It seems obvious that a ball, which deeper indents the bottom surface comes into contact with a larger area of the intermediate grease layer. So the more grease is contacted the more of it is dragged up high. Consequently, a larger amount of grease will inevitably result in longer filaments. This effect may also be corroborated by ensuing tests not discussed in this work in further detail, where thicker grease layers resulted in longer filaments.

Some impact marks are displayed in figure 5.43 for micrographic evaluation. This figure depicts representative marks created with all sixteen greases sorted by thickeners in columns and by base oils in rows. Although the marks may appear as if they protruded from the surface, in reality they still intrude into it. Reasons for this optical appearance are found in the differential interference contrast method of the microscope, which creates three dimensional reliefs of topographical inaccuracy. These images still exhibit very clearly that the surfaces are indented to differently severe degrees with mostly cone-shaped imprints. Referencing tests with the same test setup and no applied grease layer on the impacted metal surface resulted in completely spherical imprints. It is presumed that this cone shape results from the pressure distribution underneath the spherical surface of the ball. If a steel ball hits the surface it is presumed to run through several different phases. In phase one it comes into contact with the grease and displaces it when continuing any further. This is possible because, in the prevailing velocity of the ball and its resulting high shear rate, the grease is most likely to adopt liquid properties. Even though the ball displaces the grease upon emersion it will not completely replace it. Instead, the ball will rather push a local region of grease in front of it down to the metal surface. Once the ball comes very close to the metal surface the second phase is initiated, in which the grease can not yield any further in downward direction and it will try to yield in horizontal directions. This causes a high increase in pressure, which will eventually plastically deform the metal surface of the impacted plate. In

5. Experimental analysis

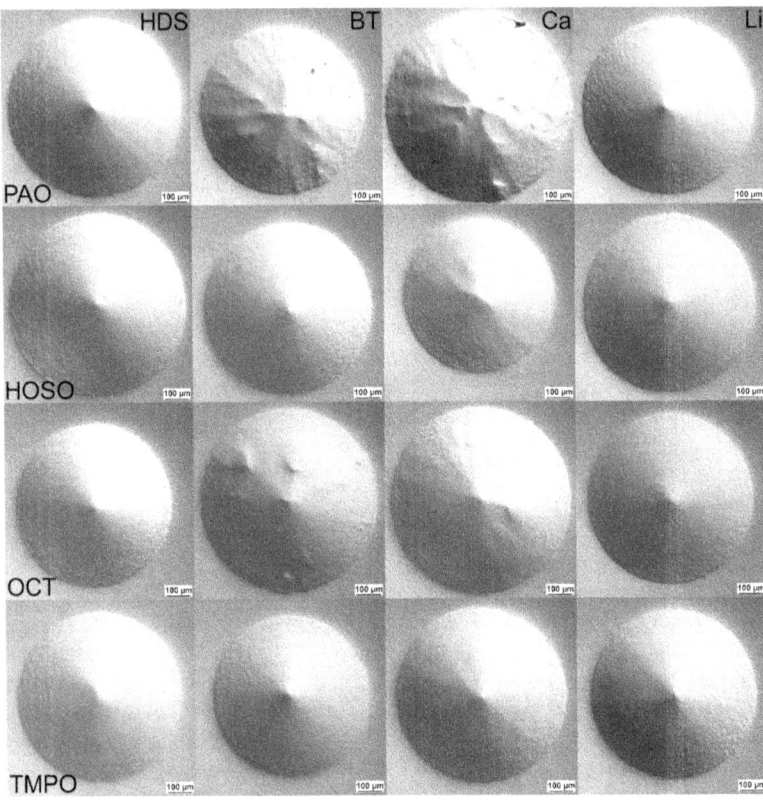

Figure 5.43.: Indentation marks created in bouncing ball tests, sorted by thickeners in columns and base oils in rows

the third phase further indentation of the plate surface by the spherical body will even seal the space in the central region of the contact and consequently result in even higher pressures. These phases will progress by continually sealing portions of the grease that were previously contacted and thus increasing their prevailing pressures. Moreover, all previously sealed regions will encounter a further increase in pressure until all of the impact energy is consumed by deformation. This way the central contacting area of the impact will experience the highest pressures and these typical cone shapes are created.

Next to the cone-shaped marks, some greases tend to create rugged, fissured imprints with concentric ridges. It is presumed that these imprints are created in a similar manner as it worked with the cones. The only difference is that in this process the thickeners locally solidified due to high pressures, which resulted in local pressure peaks and a disturbance of the aforementioned sealing of local pressure spaces.

The propensity of a grease to create these fissured, ridged imprints according to figure 5.43 seems to depend on the applied thickener type but also on the base oil. Except for HDS-PAO, in which it only occurs with a slight tendency, none of the other HDS-thickened greases show this effect. In BT-greases, on the other hand, this effect appears severely within the greases based on low-polarity oils PAO and OCT and is not found with highly polar oil-based BT-greases. None of the other greases grouped by thickeners exhibit this effect as severely as do the Ca-greases. Here too, the greases based on low polarity oils PAO and OCT more than those based on high polarity oils support the formation of highly fissured and ridged imprints. In the group of Li-greases this effect is only slightly detected with the use of TMPO as base oil.

Figure 5.43 also highlights how much the fissure and ridge forming behavior of the groups of BT- and Ca-greases is influenced by the formulated base oils resulting in a higher propensity for all greases based on low polarity oils.

To elucidate this effect one must, once again, take the microstructure of these two thickeners and the mutual interrelation of the base oils and thickeners into consideration. Local solidification was used to elucidate this effect in the afore-discussed. Now, this solidification is presumed to occur by base oil removal from the thickener in a squeezing process evoked by high pressures. It is furthermore inferred that this squeezing eventuates more likely when the enclosing base oil exerts less attraction on the thickener, which is the case with all low polarity oils. Also the size and shape of the single thickener structures seem to influence this effect. In clay thickeners it only takes place in combination with the larger bentonite particles—here the reason my be found in the platelet structure, which is presumed to favor the removal of the base oil. If platelets are squeezed against each other and are thus aligned in a parallel position they leave less space for the surrounding oil and more likely tend to extract it. If, however, the platelets are surrounded by a high polarity oil, which exerts stronger attraction to the platelets the whole structure is stabilized and the platelets are prevented from aligning parallel to each other. Consequently this squeezing effect is suppressed. In the soap thickeners this effect only happens with shorter calcium fibers, therefore it is presumed that longer lithium soap fibers help to stabilize the structure in shock impacts regardless of

5. Experimental analysis

the polarity of the applied base oil and its aforementioned consequent influence on entanglement.

Figure 5.43 also makes clear that most interestingly the original small scale-topography of the surfaces resulting from the grinding and polishing machining method remained unaffected by the indentation of the ball impacts. This effect occurs because the maximum Hertzian stress in a sphere on plane contact according to equivalent stress hypothesis is located far beneath the surface. This way only regions underneath the surface yield this stress by plastic deformation while the polishing marks on top of the surface remain unaffected. Moreover, most metallic surfaces are covered with a layer of extremely hard oxides.

The found effects may be extrapolated to real life applications, although no reasonable machine application would demand stability in such high stresses combined with these enormous shear rates. But still the results help to bring light into the processes, which occur in high-speed deformations. If greases behave in the described manner in bouncing ball tests the found results suggest that similar effects may occur in real applications such as high-speed roller bearings.

6. Main conclusions

The main findings of this work are found in the tribological and rheological behavior of the examined lubricating greases. The influence of base oils of different polarities, which is associated with bio-lubricating greases, highly affects the grease properties in the lubricated gap. The polarity dependent frictional and wear behavior postulated in the models in section 3 has been confirmed by the experimental results as the main tribological influence. This was found especially relevant in the evaluation of both frictional tests performed with the nano tribometer and by the rheometric investigations particularly in rotational mode.

Wear results of tribometer tests performed with pure base oils confirmed the model created by Massmann [51] pertaining to surface protection based on different oil polarities as defined in section 3.1.1. In further disquisition, all effects pertaining to this model will be referred to as MASSMANN-EFFECT. Other tribological results have revealed various effects that are not covered by this model alone. It was found, for example, that the interplay of thickener and base oil, depending on the material combination, is able to reverse MASSMANN-EFFECT. In summary, it can be stated that the following effects were found in the tribological investigation. The MASSMANN-EFFECT was clearly detectable in all experiments with pure base oils. It led to a better wear protection in combination with highly polar oils in the material combination steel ball on steel disc. Whereas in the sapphire steel material combination it resulted in completely opposite behavior. The MASSMANN-EFFECT had also been found within the group of soap thickened greases, yet is was mitigated to some extent. Moreover, the group of soap thickened greases resulted in quite similar wear results as did the pure base oils. The group of clay thickened greases had completely reversed the MASSMANN-EFFECT in both material combinations. Additionally, in these groups, the wear intensity in the balls was much different from the wear intensity in the steel discs.

Rheometer tests helped to confirm the model postulated in this work pertaining to the influence of oil polarities on rheological grease characteristics as outlined in section 3.1.2. All these rheological tests had revealed differently intense polarity influences. The transient tests, i.e., revealed two different polarity effects, one solely rheologically working effect and another one that influenced the thickener percentage. The latter of which led to a much higher demand of thickener quantity in the formulation of greases based on highly polar oils, primarily in combination with clay thickeners. Moreover, in combination with clay thickeners, a higher level of activation energy was necessary for structural degradation in transient shear. This same effect has been observed in the group of Li-greases but is was not found with Ca-greases. Also, SAOS tests brought forth very interesting results that help to confirm the aforementioned models. Thus, in amplitude- as well as frequency sweeps, values of G' and G'' in the LVE range reached comparable levels within the single thickener groups. These were higher values for both, G' and G'', in

6. Main conclusions

combination with highly polar oils within the groups of HDS, Ca and Li greases but, just opposite, lower values for highly polar oils within the group of BT greases. Furthermore, it was found that the evaluation of energy densities in SAOS tests was dominated by G' values inside the LVE range. Additionally, the SAOS tests also revealed the two aforementioned influences of rheological response and thickener percentage and the evaluation of $tan\delta$ values gave reason to interpret these results from a thickener share adjusted point of view. In this context, the results had proven that greases based on highly polar oils caused higher values of $tan\delta$ within all groups of thickeners but this difference was much more intense in combination with the clay thickeners. From this fact it had been deduced that high base oil polarities led to a more plasticly dominated deformation character. The tensile tests, for the most part, confirmed the results of SAOS tests as the evaluation of normal forces reflected a similar behavior as compared with the results of G' and G'' values found with amplitude and frequency sweeps inside the LVE range. Consequently, the evaluation of energy densities led to comparable results.

Below, an attempt shall be made to combine all these effects and thereby reveal the major mechanisms of mutual interrelation. In this attempt it will be elucidated which mechanisms come to act in the partial reversal of the MASSMANN-EFFECT.

Figure 6.1 depicts effects that come to act in the examined tribosystems. It

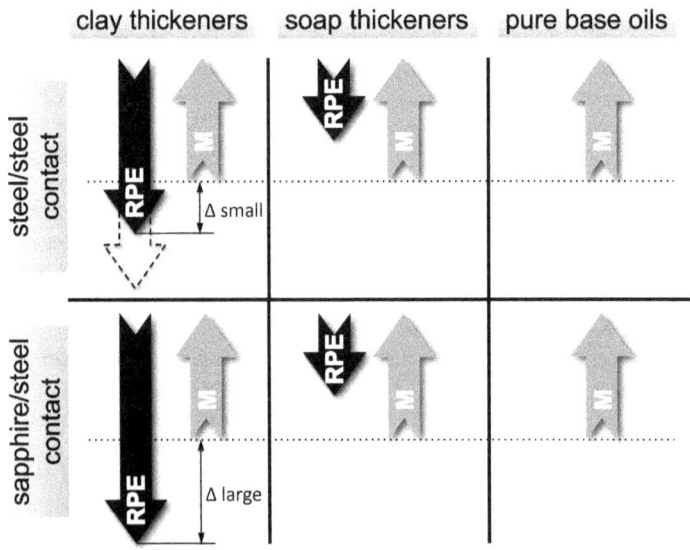

Figure 6.1.: Prevalence of effects in the investigated tribosystems—Note: M represents the Massmann effect and RPE represents the rheology polarity effect

differentiates the tribosystems according to material combination in two different rows and according to the type of thickener in three different columns. These are pure base oils for one thing (i.e. absence of thickener) but also soap and clay thickeners. This way, figure 6.1 elucidates all effects that come to act in the tribosystems displayed in figures 5.7 and 5.14, as a combination of pure oil polarity-surface effects and pure rheology-polarity effects, as well as their mutual interrelation. The former are described in model A, the latter was explained in model B, which will be referenced as Rheology-Polarity effect abbreviated by RPE in figure 6.1. All these effects are represented by arrows in figure 6.1, the length of which highlights the extent of the respective effect. The longer the arrow, the greater is the influence represented by it. The direction of the arrows indicates whether the effects support or counteract each other.

The highly polar oils, HOSO and TMPO, considered on their own, protected the surfaces in the tribocontact of steel/steel material combination significantly better than the low-polarity oils did. This protective effect is reversed in the sapphire/steel material combination, which can both be attributed to the MASSMANN-EFFECT. No other effects prevail in the tribosystems lubricated solely with pure base oils (see figure 6.1). Since the oils of all applied greases come into contact with the tribological surfaces the MASSMANN-EFFECT comes to act in every tribocontact. Therefore, the same arrows are found in all positions of figure 6.1.

The soap-thickened greases revealed a behavior similar to the pure base oils with an additionally highly reduced wear response. This highlights the fact that soap thickeners exert a significant influence on the tribological behavior. It is assumed, and has been proven [34, 130, 135], that a mix of base oil and soap fibers passes through the micro lubricated gap. When doing so, the soap fiber-base oil-mix interacting with the tribosurfaces protects the same. In case of soap-thickened greases, the MASSMANN-EFFECT also comes to act in combination with an added protective function of the soap fibers. This additional protective effect of soap fibers should theoretically be added to figure 6.1. However, this protective effect of soap fibers has been omitted since it neither supports nor counteracts the MASSMANN-EFFECT. In the proper sense, this protective effect should be added to figure 6.1 in another dimension as the displayed dimension only depicts effects that support or counteract the MASSMANN-EFFECT. Since the soap-thickened greases are also subjected to the RP-EFFECT according to model B, corresponding arrows are added to the complete intermediate column in figure 6.1. It is assumed that the RP-EFFECT counteracts the MASSMANN-EFFECT. It is reminded that rotational transient tests are those rheological tests that most closely simulate the grease behavior in the gap. Especially, a steady sliding over lubricated surfaces without manual or automatic redistribution of the grease film corresponds to the load duty of rotational rheometer tests although the former load duty is much higher. In such kind of tests it has shown that polarity influences have come to act much more severely in greases based on organically modified clay thickeners than in soap-thickened greases. Especially, the evaluation of the activation energy, necessary for the structural thickener degradation, revealed a large difference between high- and low-polarity oils-based clay-thickened greases. It is assumed that a higher measure of activation energy is necessary in the tribocontact to evoke structural degradation. By implication, this means that greases, at which less activation energy is necessary for structural degradation dissipate a larger

6. Main conclusions

amount of frictional energy to the tribological surfaces, thus leading to more wear. As the measure of activation energy was much smaller in the group of soap-thickened greases as compared with the clay greases the RP-EFFECT results in short arrows in the respective column of figure 6.1. The RP-EFFECT in the soap-thickened greases is insufficient to compensate the MASSMANN-EFFECT. Therefore the latter still dominates, although attenuated, in all soap greases.

The interplay of all found polarity influences results in a reversal of the MASSMANN-EFFECT when using the investigated inorganic non-soap thickeners in both material combinations. The findings of the tribological investigation in section 5.3 clearly revealed that clay greases based on low-polarity oils more effectively protected steel surfaces. This behavior is based on the interaction of the two established models elucidating the tribological and rheological response. As just expounded, the clay thickened greases need a higher amount of activation energy for structural degradation than those greases based on soap thickeners do. This higher amount of activation energy results in a reversal of the MASSMANN-EFFECT as displayed by long arrows representing the RP-EFFECT in figure 6.1. This reversal is more intense in the sapphire/steel than in a steel/steel material combination as shown in section 5.3, which is depicted by arrows of different length in figure 6.1.

In addition to the reversal of the MASSMANN-EFFECT the inorganic clay thickened greases also caused a substantially more aggressive wear behavior. As outlined in section 5.3, this is caused by the extreme hardness of the clay thickener particles. Viewed in this light, one should also add another arrow pertaining to the kind of thickener used. This was omitted for the same reasons the protective effect of soap thickener was not added.

In conclusion, it can be said that the presumptions and the models postulated in section 3 were validated by the experimental results performed.

Conclusiones principales

Las conclusiones más importantes de este trabajo derivan del comportamiento tribológico y reológico que presentan las grasas lubricantes estudiadas. La influencia de los aceites base de diferente polaridad, asociados al carácter biodegradable de las grasas, afecta en gran medida a las propiedades de las mismas en un contacto lubricado. La dependencia de la polaridad sobre los comportamientos en la fricción y el desgaste postulada en los modelos propuestos en la sección 3 ha sido confirmada por los resultados experimentales como la principal influencia tribológica. Esto adquiere una especial relevancia en la evaluación de los ensayos de fricción llevados a cabo con el nano-tribómetro y en los ensayos reológicos, particularmente en modo rotacional.

Los resultados de desgaste obtenidos en el tribómetro con aceites puros confirmaron el modelo propuesto por Massmann [51] en relación a la protección de superficies, basada en la polaridad de los aceites, como se ha definido en la sección 3. De aquí en adelante, todos los efectos relacionados con este modelo se referenciarán como EFECTO MASSMANN. Otros resultados tribológicos han revelado varios efectos que no están contemplados por este modelo. Se ha encontrado, por ejemplo, que la interacción entre el espesante y el aceite base, dependiendo de la combinación de materiales, puede invertir el EFECTO MASSMANN. En resumen, se puede establecer que los siguientes efectos fueron encontrados en la investigación tribológica realizada. El EFECTO MASSMANN fue claramente detectado en todos los ensayos con aceites puros. Esto condujo a una mejor protección al desgaste cuando se usan aceites altamente polares en la combinación de materiales de bola de acero-sobre-disco de acero. Sin embargo en la combinación de materiales zafiro-acero, resultó un comportamiento completamente contrario. También se ha encontrado el EFECTO MASSMANN dentro del grupo de espesantes jabonosos, pero fue mitigado en cierta medida. Además, el grupo de grasas que contienen espesantes jabonosos generó resultados de desgaste similares a los que mostraron los aceites puros. El grupo de grasas que contienen espesantes derivados de arcillas mostraron un efecto contrario al de Massmann en ambas combinaciones de materiales. Además, en estas grasas, la magnitud del desgaste en las bolas fue muy diferente respecto al desgaste de los discos de acero.

Los ensayos reológicos ayudaron a confirmar el modelo postulado en este trabajo referente a la influencia de las polaridades de los aceites sobre las propiedades reológicas de las grasas, tal y como se muestra en la sección 3.1.2. Los ensayos reológicos revelan la influencia de la polaridad de forma diferente. Por ejemplo, en los ensayos transitorios, se detectan dos efectos de polaridad diferentes, uno exclusivamente relacionado con la respuesta reológica, resultado de la interacción aceite-espesante, y otro en el que influye el porcentaje de espesante. éste último conlleva a una mayor demanda en la cantidad de espesante necesaria para la formulación de grasas basadas en aceites altamente polares, principalmente en la combinación con espesantes de arcilla. Además, con espesantes derivados de arcilla, se requiere una mayor energía de activación para la degradación estructural en ensayos transitorios. Este mismo efecto se observó en el grupo de las grasas de litio pero no se encontró para las de calcio. Además, los ensayos oscilatorios generaron resultados interesantes que ayudaron a confirmar los modelos mencionados anteriormente. Así pues, tanto en los barridos de amplitud como en los de frecuencia, los valores de G' y G'' en el rango viscoelástico lineal alcanzaron niveles comparables dentro de cada grupo de grasas que contienen el mismo espesante. Estos valores fueron mayores, tanto los de G' como los de G'', en combinaciones con aceites altamente polares dentro de los grupos de grasas espesadas con HDS, calcio y litio pero, de forma opuesta, se obtuvieron menores valores para aceites altamente polares dentro del grupo de grasas de BT. Además, se encontró que en la evaluación de las densidades de energía en ensayos oscilatorios, los valores de G' dentro del rango viscoelástico lineal son predominantes. Asimismo, los ensayos oscilatorios también pusieron de manifiesto las dos influencias anteriormente mencionadas, es decir la respuesta reológica y el porcentaje de espesante. La evaluación de la tangente de pérdidas proporciona una interpretación de estos resultados una vez aislada la influencia de la concentración de espesante. En este contexto, las grasas basadas en aceites altamente polares generaron mayores valores de la tangente de pérdidas en cada grupo de espesante, pero esta diferencia fue mucho más intensa en combinación con espesantes de arcilla. A partir de este hecho, se deduce que una alta polaridad del aceite base da lugar a una deformación plástica dominante. Los ensayos de tensión extensional, en su mayor parte, confirmaron los resultados obtenidos en los ensayos oscilatorios, ya que de la evaluación de la magnitud de la fuerza normal se deduce un comportamiento similar a los encontrados en G' y G'' en los barridos de amplitud y frecuencia dentro del rango viscoelástico lineal. Consecuentemente, la evaluación de las densidades de energía en ambos ensayos proporcionó resultados comparables.

A continuación, se expone un intento de combinar todos estos efectos y, de este modo, dilucidar los mecanismos más importantes de interrelación mutua. En este intento se esclarecerán qué mecanismos actúan en la inversión parcial del EFECTO MASSMANN.

La figura 6.2 recoge los efectos que actúan en los sistemas tribológicos examinados. Figure 6.2 depicts effects that come to act in the examined tribosystems. Se ha hecho una diferenciación de los sistemas tribológicos en función de la combinación de materiales en dos filas diferentes y en función del tipo de espesante en tres columnas diferentes. En éstas aparecen los aceites puros por un lado (en ausencia de espesante) y también las grasas espesadas con jabones metálicos y derivados de arcilla. De esta forma, la figura 6.2 recoge y aclara todos los efectos que actúan en los sistemas tribológicos mostrados en la figuras 5.7 y 5.14, como una combinación de los efectos superficiales de la polaridad de los aceites puros y los efectos de polaridad en la Reología de las grasas y, también, sus interrelaciones mutuas. Los primeros son descritos en el modelo A, y el segundo se explica según el modelo B, el cual se nombrará como efecto de la polaridad sobre la Reología, abreviado como RPE en la figura 6.2. Todos estos efectos son representados por flechas en la figura 6.2, la longitud de las cuales resaltan la magnitud de cada uno. Conforme más larga sea

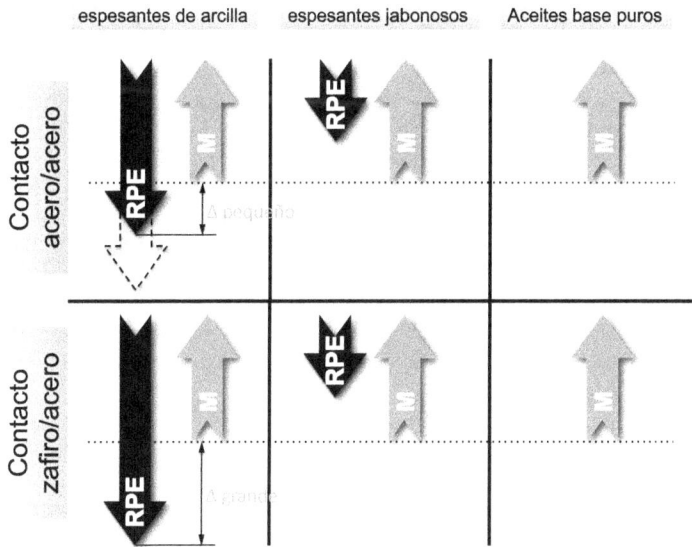

Figure 6.2.: Predominio de efectos en los sistemas tribológicos estudiados – Nota: M representa el efecto Massmann y RPE representa el efecto de la polaridad en la reología

la flecha, mayor es la influencia del efecto representado. La dirección de las flechas indica si los efectos se favorecen o contrarrestan mutuamente.

Los aceites puros que son altamente polares, HOSO y TMPO, protegieron las superficies en un contacto tribológico con combinación acero/acero significativamente mejor que lo hicieron los aceites de baja polaridad. Este efecto protector se invierte en la combinación de materiales zafiro/acero, lo cual puede atribuirse en ambos casos al EFECTO MASSMANN. No se imponen otros efectos en los sistemas tribológicos lubricados únicamente por aceites puros (ver figura 6.2). Ya que los aceites de todas las grasas aplicadas se ponen en contacto con las superficies tribológicas, el EFECTO MASSMANN actúa siempre en cada contacto tribológico. Así pues, las mismas flechas se encuentran en todas las posiciones de la figura 6.2.

Las grasas espesadas con jabones mostraron un comportamiento similar a los aceites puros, con una reducción adicional del desgaste. Esto destaca el hecho de que los espesantes jabonosos ejercen una influencia significativa sobre el comportamiento tribológico. Se ha asumido, y ha sido probado [34, 130, 135]], que una mezcla de aceites base y fibras de jabones pasan a través de un micro-contacto lubricado. Cuando ocurre esto, la mezcla aceite-fibra que interacciona con las superficies tribológicas protegen a las mismas. En el caso de grasas espesadas con jabones, el EFECTO MASSMANN actúa también en combinación con una función añadida de protección de las fibras del jabón. Este efecto protector adicional de las fibras de jabón debería ser añadido teóricamente en la figura 6.2. Sin embargo, este efecto protector de las fibras de jabón ha sido omitido ya que ni confirman ni contrarrestan el EFECTO MASSMANN. En este sentido, este efecto protector debería añadirse a la figura 6.2 en otra dimensión, ya que la dimensión mostrada sólo representa efectos que confirman o contrarrestan el EFECTO MASSMANN. Ya que las grasas espesadas con jabones se encuentran sometidas al EFECTO RP según el modelo B, las flechas correspondientes se han añadido a la columna intermedia en la figura 6.2. Se ha asumido que el EFECTO RP contrarresta el EFECTO MASSMANN. Se recuerda que los ensayos rotacionales transitorios son aquellos ensayos reológicos que simulan de forma más realista el comportamiento de la grasa en un contacto tribológico. El continuo deslizamiento sobre las superficies lubricadas sin redistribución manual o automática de la película de grasa se corresponde con la de un ensayo transitorio, aunque la carga en un contacto real es mucho mayor. En este tipo de ensayos se ha mostrado que la influencia de la polaridad actúa de forma mucho más severa en grasas basadas en arcillas modificadas orgánicamente que en grasas espesadas con jabón. Especialmente, la evaluación de la energía de activación, necesaria para la degradación estructural del espesante, reveló grandes diferencias entre aceites con alta y baja polaridad en grasas espesadas con arcillas. Se asume que es necesaria una mayor energía de activación en un contacto tribológico para provocar la degradación estructural del espesante. Por consiguiente, esto significa que aquellas grasas con una menor energía de activación necesaria para la degradación estructural disipan una cantidad mayor de energía de fricción hacia las superficies tribológicas, generando así mayor desgaste. Como la medida de la energía de activación fue mucho menor en el grupo de las grasas espesadas con jabones en comparación con las de arcilla, el EFECTO RP se representa con flechas más cortas en la columna correspondiente de la figura 6.2. El EFECTO RP en las grasas espesadas con jabones es insuficiente para compensar el EFECTO MASSMANN. Así pues, éste último todavía domina, aunque atenuado, en todas las grasas espesadas con jabón.

La interacción de todas las influencias de la polaridad de los aceites base resulta en una inversión del EFECTO MASSMANN cuando se usaron espesantes inorgánicos no jabonosos en ambas combinaciones de materiales. Los resultados de la investigación tribológica en la sección 5.3 revelaron claramente que las grasas de arcilla basadas en aceites poco polares protegen de manera más efectiva las superficies de acero. Este comportamiento se basa en la interacción de los dos modelos establecidos que clarifican la respuesta tribológica y reológica. Como se ha expuesto, las grasas espesadas con derivados de arcilla necesitan una mayor cantidad de energía de activación para la degradación estructural que aquellas basadas en espesantes jabonosos. Esta mayor cantidad de energía de activación genera una inversión del EFECTO MASSMANN como se muestra a través de flechas largas que representan el EFECTO RP en la figura 6.2. Esta inversión es más intensa en la combinación de materiales zafiro/acero que en la acero/acero, tal y como se mostró en la sección 5.3, la cual es representada por flechas de diferente longitud en la figura 6.2. Además de la inversión del EFECTO MASSMANN, las grasas espesadas con arcillas inorgánicas también generaron un desgaste de las superficies sustancialmente más agresivo. Tal y como se explica en la sección 5.3, esto es causado por la dureza extrema de las partículas del espesante de arcilla. Visto de esta forma, se debería añadir otra flecha que haga referencia al tipo de espesante usado. Esto fue omitido por las mismas razones por las cuales el efecto protector de los espesantes jabonosos no fue añadido.

En conclusión, se puede decir que las suposiciones y los modelos postulados en la sección 3 fueron validados por los resultados experimentales llevados a cabo.

Bibliography

[1] Bentley, R.W.: Global oil & gas depletion: an overview, Energy Policy 30, 189–205 (2002)

[2] Rogner, H.H.: An Assessment of World Hydrocarbon Resources, Annual Review of Energy and the Environment 22, 217–262 (1997)

[3] Wilson, B.: Lubricants and functional fluids from renewable sources, Industrial lubrication and tribology 50, 6–15 (1998)

[4] Willing, A.: Lubricants based on renewable resources - an environmentally compatible alternative to mineral oil products, Chemosphere 43, 89–98 (2001)

[5] Mercurio, P., Burns, K., Negri, A.: Testing the ecotoxicology of vegetable versus mineral based lubricating oils: 1. Degradation rates using tropical marine microbes, Environmental Pollution 129, 165–173 (2004)

[6] Wagner, H., Luther, R., Mang, T.: Lubricant base fluids based on renewable raw materials: Their catalytic manufacture and modification, Applied Catalysis A: General 221, 429–442 (2001)

[7] Dresel, W.: Biologically degradable lubricating greases based on industrial crops, Industrial Crops and Products 2, 281–288 (1994)

[8] Kuhn, E.: Description of the energy level of tribologically stressed greases, Wear 188, 138–141 (1995)

[9] Kuhn, E., Balan, C.: Experimental procedure for the evaluation of the friction energy of lubricating greases, Wear 209, 237–240 (1997)

[10] Kuhn, E., Glüsing, H.: Systematic effects on the precision of penetration readings - a contribution to the rheology of lubricating greases, Eurogrease 17–28 (1997)

[11] Kuhn, E.: Experimental grease investigations from an energy point of view, Industrial Lubrication and Tribology 51, 246–251 (1999)

[12] Kuhn, E.: An algorithm to estimate the friction energy of a grease lubricated contact, Industrial Lubrication and Tribology 52, 174–7 (2000)

[13] Kuhn, E., Schmidt, T.: Investigation into the cohesion behaviour of lubricating greases with a new pendulum tribometer, Eurogrease 15–18 (2002)

[14] Kuhn, E., Holweger, W.: Topography and tribological behaviour of lubricating greases-an experimental study, Industrial Lubrication and Tribology 56, 14–18 (2004)

[15] Kuhn, E.: Deformation tests with model greases by using a rheometer, Eurogrease 6–10 (2005)

[16] Kuhn, E.: Irrevesible eigenschaftsänderung durch reibungsbeanspruchung bei schmierfetten - rheologischer verschleiß, Tribologie und Schmierungstechnik 53, 27–29 (2006)

[17] Kuhn, E.: Description of the energy level of grease lubricated contacts, in AITC-AIT, 5th International Conference of Tribology, vol. 20 (2006)

[18] Kuhn, E.: Über die Unmöglichkeit eines verschleißlosen tribologischen Prozesses, Tribologie und Schmierungstechnik 54, 10–12 (2007)

[19] Kuhn, E.: Influence of the soap content of lubricating greases on the tribological process, Eurogrease 7–11 (2007)

[20] Kuhn, E.: Analysis of a grease-lubricated contact from an energy point of view, International Journal of Materials and Product Technology 38, 5–15 (2010)

[21] Kuhn, E., Delgado, M.: Description of the structural degradation of lubricating greases as a reaction of the tribological system, Eurogrease 12–15 (2010)

[22] Kuhn, E.: Der Schmierstoffverschleiß, Tribologie und Schmierungstechnik 58, 32–35 (2011)

[23] Kuhn, E.: Investigation of the structural degradation of lubrcating greases due to tribological stress, in W.J. Bartz (ed.), 18th International Colloquium of Tribology, Esslingen, Technische Akademie Esslingen (2012)

[24] Kuhn, E.: Zur Tribologie der Schmierfette, expert Verlag, Renningen (2009)

[25] Lugt, P.M.: A review on grease lubrication in rolling bearings, Tribology Transactions 52, 470–480 (2009)

[26] Venner, C., van Zoelen, M., Lugt, P.: Thin layer flow and film decay modeling for grease lubricated rolling bearings, Tribology International 47, 175–187 (2012)

[27] Balan, C., Franco, J.M.: Influence of the geometry on the transient and steady flow of lubricating greases, Tribology Transactions 44, 53–58 (2001)

[28] Franco, J., Delgado, M., Valencia, C., Sánchez, M., Gallegos, C.: Mixing rheometry for studying the manufacture of lubricating greases, Chemical Engineering Science 60, 2409–2418 (2005)

[29] Martin-Alfonso, J., Valencia, C., Sanchez, M., Franco, J., Gallegos, C.: Development of new lubricating grease formulations using recycled LDPE as rheology modifier additive, European Polymer Journal 43, 139 – 149 (2007)

[30] Delgado, M., Franco, J., Valencia, C., Kuhn, E., Gallegos, C.: Transient shear flow of model lithium lubricating greases, Mechanics of Time-Dependent Materials 13, 63–80 (2009)

[31] Delgado, M., Valencia, C., Sanchez, M., Franco, J., Gallegos, C.: Thermorheological behaviour of a lithium lubricating grease, Tribology Letters 23, 47–54 (2006)

[32] Yokouchi, A., Yamamoto, Y.: Influence of Soap Fiber Structure on Frictional Property of Lithium Soap Grease, ASME Conference Proceedings 2007, 129–131 (2007)

[33] Rong-Hua, J.: Effects of the composition and fibrous texture of lithium soap grease on wear and friction, Tribology International 18, 121–124 (1985)

[34] Hurley, S., Cann, P.: Grease Composition and Film Thickness in Rolling Contacts, NLGI Spokesman 63, 12–22 (1999)

[35] Cousseau, T., Graça, B., Campos, A., Seabra, J.: Influence of grease formulation on thrust bearings power loss, Journal of Engineering Tribology 224, 935–946 (2010)

[36] Boner, C.: Manufacture and application of lubricating greases, Reinhold (1954)

[37] Bondi, A., Fraser, H., Abrams, S., Moore, R., R., Smith, G. H., A., White, A., E., Wilson, B., J., Bondi, A, A., Stross, F., Peterson, D., W., et al.: Developments in the field of soda base greases, in 3rd World Petroleum Congress (1951)

[38] Farrington, B.B.: The fine structure of lubricating greases, Annals of the New York Academy of Sciences 53, 979–986 (1951)

[39] Fagan, G.L.: The 2005 grease production survey report, NLGI Spokesman-Including NLGI Annual Meeting-National Lubricating Grease Institute 70, 25–48 (2007)

[40] Bergseth, H.: Viskosität einer Na^+-Montmorillonit-Suspension nach Verdünnungen und Salzzusätzen, Colloid & Polymer Science 189, 63–66 (1963)

[41] Goerz, T.: Gel-und Bentonitfette Zusammensetzung -Eigenschaften, Tribologie und Schmierungstechnik 56, 30–34 (2009)

[42] Alther, G.R.: The effect of the exchangeable cations on the physico-chemical properties of wyoming bentonites, Applied Clay Science 1, 273–284 (1986)

[43] Bergaya, F., Lagaly, G.: Surface modification of clay minerals, Applied Clay Science 19, 1–3 (2001)

Bibliography

[44] Pogosyan, A., Martirosyan, T.: Tribological properties of bentonite thickener-containing greases, Journal of Friction and Wear 29, 205–209 (2008)

[45] Chtourou, M., Frikha, M., Trabelsi, M.: Modified smectitic Tunisian clays used in the formulation of high performance lubricating greases, Applied clay science 32, 210–216 (2006)

[46] Kohashi, H.: Application of fatty acid esters for lubricating oil, in World Conference on Oleochemicals Into the 21st Century: Proceedings, American Oil Chemists Society, 243 (1991)

[47] Alther, G.: Organically modified clay removes oil from water, Waste Management 15, 623–628 (1995)

[48] Dekra Betriebsstoffliste 2013, trans aktuell spezial, EuroTransportMedia Verlag (2013)

[49] Bartz, W.: Ökologische und ökonomische Aspekte bei Schmierstoffen.: Industriehygiene und Produktsicherheit - Anwendung und Entsorgung., Kontakt u. Studium, Expert-Verlag GmbH (2001)

[50] Webber, M.: Petroleum 76 (1945)

[51] Massmann, T.: Wirkmechanismen additivierter Schmierstoffe in der Kaltumformung, Shaker (2007)

[52] Bartz, W.: Additive für Schmierstoffe, Kontakt u. Studium, Expert-Verlag GmbH (1994)

[53] Mang, T., Dresel, W.: Lubricants and Lubrication, Wiley (2007)

[54] Aguilar, G., Mazzamaro, G., Rasberger, M.: Oxidative degradation and stabilisation of mineral oil-based lubricants, in R.M. Mortier, M.F. Fox, S.T. Orszulik (eds.), Chemistry and Technology of Lubricants, Springer Netherlands, 107–152 (2010)

[55] Paolino, C.: Antioxidants, in J. Lutz (ed.), Thermoplastic Polymer Additives: Theory and Practice, M. Dekker, Plastics engineering (1989)

[56] Reyes-Gavilan, J., Odorisio, P.: A review of the mechanisms of action of antioxidants, metal deactivators, and corrosion inhibitors, NLGI spokesman 64, 22–33 (2001)

[57] Rudnick, L.: Lubricant Additives: Chemistry and Applications, Second Edition, Chemical Industries, Taylor & Francis (2010)

[58] Czichos, H., Hennecke, M.: Hütte: Das Ingenieurwissen, Springer Verlag (2007)

[59] gGmbH, R.: Basic criteria for award of the environmental label - readily biodegradable lubricants and forming oils ral-uz 64, Tech. Rep., RAL gGmbH (2011)

[60] Sanchez, R., Franco, J., Delgado, M., Valencia, C., Gallegos, C.: Thermal and mechanical characterization of cellulosic derivatives-based oleogels potentially applicable as bio-lubricating greases: Influence of ethyl cellulose molecular weight, Carbohydrate Polymers 83, 151–158 (2011)

[61] Sanchez, R., Stringari, G., Franco, J., Valencia, C., Gallegos, C.: Use of chitin, chitosan and acylated derivatives as thickener agents of vegetable oils for bio-lubricant applications, Carbohydrate Polymers 85, 705–714 (2011)

[62] Sanchez, R., Franco, J., Delgado, M., Valencia, C., Gallegos, C.: Rheological and mechanical properties of oleogels based on castor oil and cellulosic derivatives potentially applicable as bio-lubricating greases: Influence of cellulosic derivatives concentration ratio, Journal of Industrial and Engineering Chemistry 17, 705–711 (2011)

[63] Sanchez, R., Franco, J., Kuhn, E., Fiedler, M.: Tribological characterization of green lubricating greases formulated with castor oil and different biogenic thickener agents: a comparative experimental study, Industrial Lubrication and Tribology 63, 446–452 (2011)

[64] Stachowiak, G., Batchelor, A.: Engineering Tribology, Engineering Tribology, Elsevier Science (2011)

[65] Owens, D., Wendt, R.: Estimation of the surface free energy of polymers, Journal of Applied Polymer Science 13, 1741–1747 (1969)

[66] Bondi, A.: van der Waals Volumes and Radii, The Journal of Physical Chemistry 68, 441–451 (1964)

[67] Habenicht, G.: Kleben: Grundlagen, Technologien, Anwendungen, VDI-Buch, Springer (2008)

[68] Lawrence, A.S.C.: The peptisation of aqueous soap solutions, Trans Faraday Soc 33, 325–330 (1937)

[69] McBain, J.W., Mysels, K.J., Smith, G.H.: Studies of aluminium soaps. vii. aluminium soaps in hydrocarbons. the gels and jellies and transformations between them, Trans Faraday Soc 42, B173–B180 (1946)

[70] Schwuger, M., Findenegg, G.: Lehrbuch der Grenzflächenchemie, Wiley VCH Verlag GmbH (2001)

[71] Gohar, R., Rahnejat, H.: Fundamentals of tribology, Imperial College Press (2008)

[72] Bartz, W.J.: Zur Geschichte der Tribologie, vol. 1, expert verlag (1988)

Bibliography

[73] Koner, Guhl: Leben der Griechen und Römer, Engelmann (1893)

[74] Fleischer, G., Gröger, H., Thum, H.: Verschleiß und Zuverlässigkeit, VEB Verlag Technik, Berlin (1980)

[75] Czichos, H.: Tribologie-Handbuch, Vieweg und Teubner, Wiesbaden (2010)

[76] Schmaltz, G.: Techische Oberflächenkunde. Feingestalt und Eigenschaften von Grenzflächen technischer Körper, insbesondere der Maschinenteile, Springer (1936)

[77] Fleischer, G., Dzimko, M., Bosse, H., Buchmann, R., Winkelmann, U.: Berechnung der Reibung auf energetischer Grundlage, Wissenschaftliche Zeitschrift der Universität Magdeburg (1982)

[78] Holweger, W.: Spectral density of lubricants in tribological contacts: towards a new sight of lubricants design for life, Trib Coll Ecole Polytechnique Federal de Lausanne (1999)

[79] Fleischer, G.: Zum energetischen Niveau von Reibpaarungen, Schmierungstechnik 16 (1985)

[80] Fleischer, G.: Zur energetik der reibung, Wissenschaftliche Zeitschrift der Universität Magdeburg (1990)

[81] Fleischer, G.: Die Tross'schen Erkenntnisse aus heutige Sicht, in E. Kuhn (ed.), 3. Arnold Tross Kolloquium, Hamburg University of Applied Sciences, Shaker Verlag, vol. 3 (2007)

[82] Winkelmann, U.: Grundlagen zur energetischen Bestimmung von Reibungskenngrößen bei Festkörperreibung metallischer Gleitpaarungen, Ph.D. thesis, Otto von Guericke Universität Magdeburg (1981)

[83] Rühle, F.: Experimentelle Ermittlung der Eigenschaften oberflächennaher Stoffbereiche bei unterschiedlichen Fertigungsvefahren, in E. Kuhn (ed.), 3. Arnold Tross Kolloquium, Hamburg University of Applied Sciences, Shaker Verlag, vol. 3 (2007)

[84] W., T.: Reibung und Verschleiß an Reibpaarungen mit Gleitbewegung beim Wirken abrasiver Teilchen, Ph.D. thesis, Otto von Guericke Universität Magdeburg (1979)

[85] Patzelt, B.: Simulation einer stoßartigen Beanspruchung mit dem Pendelfurcher, Ph.D. thesis, Otto von Guericke Universität Magdeburg (1995)

[86] Czarny, R.: Effects of changes in grease structure on sliding friction, Industrial Lubrication and Tribology 47, 3–7 (1995)

[87] Zum Gahr, K.H.: Microstructure and wear of materials, vol. 10, Elsevier Science Limited (1987)

[88] Franco, J., Delgado, M., Valencia, C.: Combined oxidative shear resistance of castor oil based greases, in E. Kuhn (ed.), 3. Arnold Tross Kolloquium, Shaker Verlag, vol. 3 (2006)

[89] Kuhn, E.: Inherent tribo-system response to optimise the process conditions, in 8th Arnold-Tross-Kolloquium (2012)

[90] Abdel-Aal, H.: Wear and irreversible entropy generation in dry sliding, Annals Dunarea de Jos of Galati (Tribology) 34–44 (2006)

[91] Doelling, K.L., Ling, F.F., Bryant, M.D., Heilman, B.P.: An experimental study of the correlation between wear and entropy flow in machinery components, Journal of Applied Physics 88, 2999–3003 (2000)

[92] Viafara, C., Sinatora, A.: Thermodynamic approaches in sliding wear: a review, International Journal of Materials and Product Technology 38, 93–116 (2010)

[93] Ling, F., Bryant, M., Doelling, K.: On irreversible thermodynamics for wear prediction, Wear 253, 1165–1172 (2002)

[94] Delgado, M., Franco, J., Kuhn, E.: Effect of rheological behaviour of lithium greases on the friction process, Industrial Lubrication and Tribology 60, 37–45 (2008)

[95] Malkin, A.: Rheology Fundamentals, Fundamental topics in rheology, Elsevier Science & Technology (1994)

[96] Ferry, J.: Viscoelastic Properties of Polymers, Wiley (1980)

[97] Mezger, T.: Das Rheologie Handbuch: Für Anwender von Rotations-und Oszillations-rheometern, Vincentz Network GmbH & Co KG, Hannovr, Germany (2007)

[98] Steffe, J.: Rheological Methods in Food Process Engineering, Freeman Press (1996)

[99] Delgado, M., Valencia, C., Sanchez, M., Franco, J., Gallegos, C.: Influence of soap concentration and oil viscosity on the rheology and microstructure of lubricating greases, Industrial and Engineering Chemistry Research 45, 1902–1910 (2006)

[100] Martin-Alfonso, J., Valencia, C., Sanchez, M., Franco, J., Gallegos, C.: Recycled and virgin ldpe as rheology modifiers of lithium lubricating greases: A comparative study, Polymer Engineering & Science 48, 1112–1119 (2008)

[101] Martin-Alfonso, J.E., Valencia, C., Sanchez, M.C., Franco, J.M.: Evaluation of thermal and rheological properties of lubricating greases modified with recycled ldpe, Tribology Transactions 55, 518–528 (2012)

[102] Bartz, W.J.: Ecotribology: Environmentally acceptable tribological practices, Tribology International 39, 728–733, 4th AIMETA International Tribology Conference (2006)

[103] Battersby, N., Ciccognani, D., Evans, M., King, D., Painter, H., Peterson, D., Starkey, M.: An inherent biodegradability test for oil products: Description and results of an international ring test, Chemosphere 38, 3219–3235 (1999)

[104] Battersby, N.S.: The biodegradability and microbial toxicity testing of lubricants - some recommendations, Chemosphere 41, 1011–1027 (2000)

[105] Lugscheider, E., Bobzin, K.: The influence on surface free energy of pvd-coatings, Surface and Coatings Technology 142-144, 755–760, proceedings of the 7th International Conference on Plasma Surface Engineering (2001)

[106] Lugscheider, E., Bobzin, K.: Wettability of pvd compound materials by lubricants, Surface and Coatings Technology 165, 51–57 (2003)

[107] Wu, S.: Calculation of interfacial tension in polymer systems, Journal of Polymer Science Part C: Polymer Symposia 34, 19–30 (1971)

[108] Fowkes, F.: Additivity of intermolecular forces at interfaces. i. determination of the contribution to surface and interfacial tensions of dispersion forces in various liquids1, The Journal of Physical Chemistry 67, 2538–2541 (1963)

[109] Fowkes, F.M.: Attractive forces at interfaces, Industrial & Engineering Chemistry 56, 40–52 (1964)

[110] Fowkes, F.M.: Donor-acceptor interactions at interfaces, The Journal of Adhesion 4, 155–159 (1972)

[111] Zainal, Isengard, H.D.: Determination of total polar material in frying oil using accelerated solvent extraction, Lipid Technology 22, 134–136 (2010)

[112] Dobarganes, M.C., Velasco, J., Dieffenbacher, A.: Determination of polar compounds, polymerized and oxidized triacylglycerols, and diacylglycerols in oils and fats, Pure and Applied Chemistry 72, 1563–1575 (2000)

[113] Knorn, B., Göbel, R., Franzke, C.: Bestimmung der Polarität gesättigter Monoglyceride durch Gaschromatographie, Food / Nahrung 21, 817–823 (1977)

[114] Rouser, G., Kritchevsky, G., Galli, C., Heller, D.: Determination of polar lipids: Quantitative column and thin-layer chromatography, Journal of the American Oil Chemists' Society 42, 215–227 (1965)

[115] Márquez-Ruiz, G., Jorge, N., Martín-Polvillo, M., Dobarganes, M.C.: Rapid, quantitative determination of polar compounds in fats and oils by solid-phase extraction and size-exclusion chromatography using monostearin as internal standard, Journal of Chromatography A 749, 55–60 (1996)

[116] Ringard-Lefebvre, C., Bochot, A., Memisoglu, E., Charon, D., Duchêne, D., Baszkin, A.: Effect of spread amphiphilic [beta]-cyclodextrins on interfacial properties of the oil/water system, Colloids and Surfaces B: Biointerfaces 25, 109–117 (2002)

[117] El-Mahrab-Robert, M., Rosilio, V., Bolzinger, M.A., Chaminade, P., Grossiord, J.L.: Assessment of oil polarity: Comparison of evaluation methods, International Journal of Pharmaceutics 348, 89–94 (2008)

[118] Kistler, S.S.: The measurement of bound water by the freezing methods, Journal of the American Chemical Society 58, 901–907 (1936)

[119] Anderson, F., Nelson, R., Farley, F.: Preparation of grease specimens for electron microscopy, NLGI Spokesman 31, 252 (1967)

[120] Brown, J., Hudson, C., Loring, L.: Lithium grease-detailed study by electron microscope, The Institute Spokesman 15, 8–17 (1952)

[121] Shuff, P., Clarke, L.: Imaging of lubricating oil insolubles by electron microscopy, Tribology International 24, 381–387 (1991)

[122] Mansot, J., Terech, P., Martin, J.: Structural investigation of lubricating greases, Colloids and Surfaces 39, 321–333 (1989)

[123] Sanchez, M., Franco, J., Valencia, C., Gallegos, C., Urquiola, F., Urchegui, R.: Atomic Force Microscopy and Thermo-Rheological Characterisation of Lubricating Greases, Tribology Letters 41, 463–470 (2011)

[124] Mortimer, C., Müller, U.: Chemie: Das Basiswissen der Chemie, Thieme (2003)

[125] Becker, G., Braun, D.: Kunststoff-Handbuch: Duroplaste, Hanser Verlag (1988)

[126] Gallay, W., Puddington, I.E.: The hydration of starch below the gelatinization temperature, Canadian Journal of Research 21b, 179–185 (1943)

[127] Delgado, M., Sanchez, M., Valencia, C., Franco, J., Gallegos, C.: Relationship among microstructure, rheology and processing of a lithium lubricating grease, Chemical Engineering Research and Design 83, 1085–1092 (2005)

[128] Moore, R.J., Cravath, A.M.: Mechanical breakdown of soap-base greases., Industrial & Engineering Chemistry 43, 2892–2897 (1951)

[129] Fiedler, M.: Complexity of tribological characterizations illustrated with poly-alpha-olefin grease, Eurogrease 1, 4–7 (2011)

[130] Cann, P., Hurley, S.: Friction Properties of Grease in EHD Lubrication, NLGI Spokesman 66, 6–15 (2002)

Bibliography

[131] Kimura, H., Imai, Y., Yamamoto, Y.: Study on Fiber Length Control for Ester-Based Lithium Soap Grease, Tribology Transactions 44, 405–410 (2001)

[132] Fiedler, M., Käfer, E.: Tribologisches Potential von Schmierfetten auf Basis biogener Grundöle, in Fachtagung der GfT 2010, GfT (2010)

[133] Fiedler, M.: Tribological Potential of Greases based on biogenous Esters, in 6th Arnold Tross Kolloquium, Shaker Verlag (2010)

[134] Peng, D., Chen, C., Kang, Y., Chang, Y., Chang, S.: Size effects of $SiO2$ nanoparticles as oil additives on tribology of lubricant, Industrial Lubrication and Tribology 62, 111–120 (2010)

[135] Cann, P., Webster, M., Doner, J., Wikstrom, V., Lugt, P.: Grease Degradation in R0F Bearing Tests, Tribology Transactions 50, 187–197 (2007)

List of Figures

2.1.	Common thickeners in lubricating greases	6
2.2.	SEM images of Fibers of various lubricating greases	7
2.3.	Size comparison between single soap fibers and various viruses and bacteria	8
2.4.	Montmorillonite layers in scanning electron microscopy	10
2.5.	Ester reaction in oleochemicals as found in [49]	11
2.6.	Basic structure of a tribological system following [74]	21
2.7.	Cross-section of a metallic surface following [76] as found in [75]	22
2.8.	Model of a real contact situation taken from [75]	23
2.9.	Model of a grease lubricated contact situation following [24]	24
2.10.	Model of sliding friction taken from [74]	25
2.11.	Model of rolling friction taken from [74]	26
2.12.	Model of drilling friction taken from [74]	26
2.13.	Model of solid state friction as taken from [74]	27
2.14.	Model of liquid state friction as taken from [24]	28
2.15.	Friction states in the Stribeck curve—taken from [24]	29
2.16.	Model of mixed state friction as taken from [24]	29
2.17.	Possible proportions of friction energy states [79]	31
2.18.	Energetic distribution and contribution of friction bodies to the tribological process [80]	31
2.19.	Classification of tribological wear following [24]	33
2.20.	Wear mechanisms according to [87] taken from [75]	35
2.21.	Model of entropy transport or entropy flow into the system	41
2.22.	Example curve of a rotational transient test	42
2.23.	Example curves of characteristic sinusoidal load in small amplitude oscillatory shear (SAOS)	44
2.24.	Example curve of an amplitude sweep rheometer test including characteristic points	45
2.25.	Example curve of a tensile test	48
3.1.	Model of oil polarity influence on tribological response	54
4.1.	Experimental setup of the nanotribometer	62
5.1.	AFM image of Li-HOSO grease	68
5.2.	AFM image of Li-TMPO grease	69
5.3.	AFM image of Li-OCT grease	70
5.4.	AFM image of BT-HOSO grease	71
5.5.	AFM image of HDS-OCT grease	72

List of Figures

5.6. Friction coefficient in contact situation sapphire ball on steel disc .. 74
5.7. Widths of wear tracks in sapphire balls and steel discs 76
5.8. HDS-OCT in material combination sapphire ball on steel disc 78
5.9. BT-HOSO in material combination sapphire ball on steel disc 79
5.10. Ca-TMPO in material combination sapphire ball on steel disc 81
5.11. Li-OCT in material combination sapphire ball on steel disc 82
5.12. Pure base oils in material combination sapphire ball on steel disc .. 83
5.13. Friction coefficient in contact situation steel ball on steel disc 85
5.14. Widths of wear tracks in steel balls and steel discs 85
5.15. HDS-PAO in material combination steel ball on steel disc 86
5.16. BT-TMPO in material combination steel ball on steel disc 87
5.17. Ca-PAO in material combination steel ball on steel disc 88
5.18. Li-HOSO in material combination steel ball on steel disc 89
5.19. Pure base oils in material combination steel ball on steel disc 90
5.20. AFM Profile of speckles and wear track—BT-grease 100 mN in material combination sapphire ball on steel disc 95
5.21. Transient flow curves of selected Li-grease samples determined at 40 °C along with the fits of the corresponding curves according to equation 2.38 [24] 99
5.22. Energy densities, $e_{rheo-rot}$, determined by integration of shear stress over time at -10 °C 100
5.23. Energy densities $e_{rheo-rot}$ determined by integration of shear stress over time at 40 °C 100
5.24. Chart of different polarity influences on structural breakdown in transient shear 102
5.25. Plot of energy densities $e_{rheo-rot}$ determined at all given temperatures 103
5.26. Original data and Arrhenius fits of greases based on highly polar HOSO and non-polar PAO 104
5.27. Activation energy E_a, in transient shear flow 105
5.28. Dataplot of G' and G'' of all 16 greases sorted by thickeners—amplitude sweep 108
5.29. Diagram of the storage modulus G' of all 16 greases sorted in groups of thickeners 109
5.30. Diagram of the loss modulus G'' of all 16 greases sorted in groups of thickeners 109
5.31. Loss tangent, $tan(\delta)$, of all 16 greases sorted in groups of thickeners—values determined in the LVE range 110
5.32. Values of $G' = G''$ of all 16 greases sorted in groups of thickeners .. 112
5.33. Deformation in % of all 16 greases sorted in groups of thickeners—values determined in the cross over point 112
5.34. Energy densities of amplitude sweep tests—LVE-range 114
5.35. Apparent rheological frictional energy densities of amplitude sweep tests—Crossover point 114
5.36. Frequency sweep dataplot of G' and G'' of all 16 greases sorted by groups of thickeners 117
5.37. G' and G'' determined with averaged values of frequency sweeps .. 118

List of Figures

5.38. Energy densities determined with frequency sweeps 119
5.39. Normal forces determined in tensile tests 120
5.40. Energy densities determined in tensile tests 121
5.41. Filament lengths measured in bouncing ball tests 124
5.42. Impact diameters measured in bouncing ball tests 125
5.43. Indentation marks created in bouncing ball tests 126

6.1. Prevalence of effects in the investigated tribosystems 130
6.2. Predominio de efectos en los sistemas tribológicos estudiados 135

A.1. HDS-OCT in material combination sapphire ball on steel disc 154
A.2. BT-PAO in material combination sapphire ball on steel disc 155
A.3. Ca-TMPO in material combination sapphire ball on steel disc 156
A.4. Li-HOSO in material combination sapphire ball on steel disc 157

A. Additional information pertaining section 5.3.1

A.1. Interferometric investigation of selected wear marks

In addition to the micrographs of wear results, which have been so far discussed, another micro analysis was performed by means of three-dimensional optical scanning with white light interferometry. For this purpose some tribosystems were selected for further investigation. The tests were repeated with equal sapphire balls and steel plates as used in the aforementioned evaluation. Results are displayed in figures A.1 to A.4. Each of these figures provides a spacial view on the respective wear marks created with the given greases in sapphire/steel contact and a surface scanning profile derived from the interferometrically taken images.

A. Additional information pertaining section 5.3.1

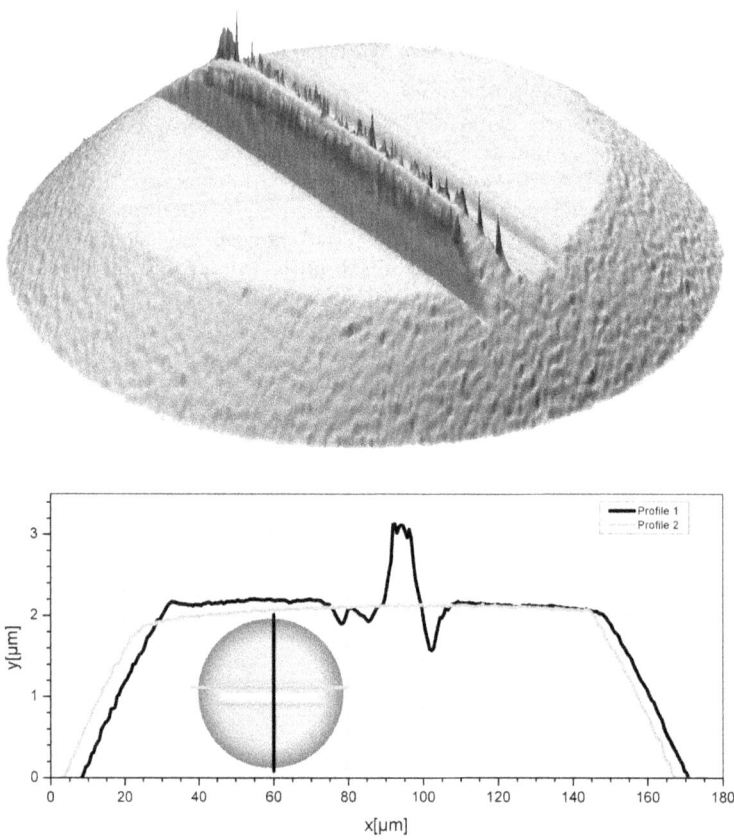

Figure A.1.: HDS-OCT in material combination sapphire ball on steel disc—spacial view created with withe light interferometry in the upper part and profiles in the lower part with profile sections as exhibited in the miniature

A.1. Interferometric investigation of selected wear marks

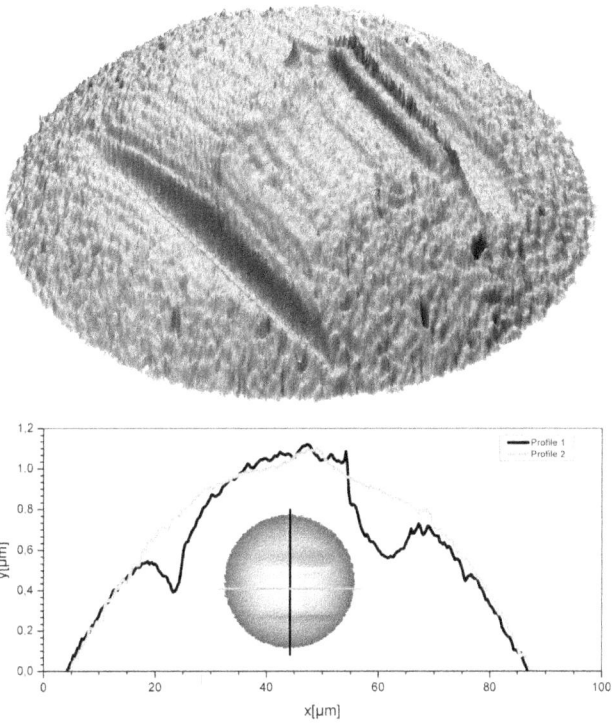

Figure A.2.: BT-PAO in material combination sapphire ball on steel disc—spacial view created with withe light interferometry in the upper part and profiles in the lower part with profile sections as exhibited in the miniature

A. Additional information pertaining section 5.3.1

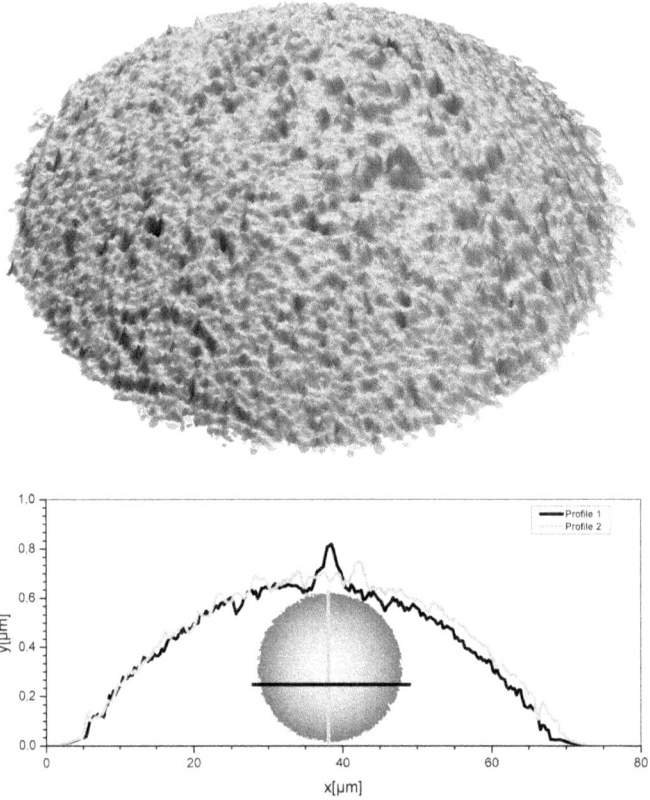

Figure A.3.: Ca-TMPO in material combination sapphire ball on steel disc—spacial view created with withe light interferometry in the upper part and profiles in the lower part with profile sections as exhibited in the miniature

A.1. Interferometric investigation of selected wear marks

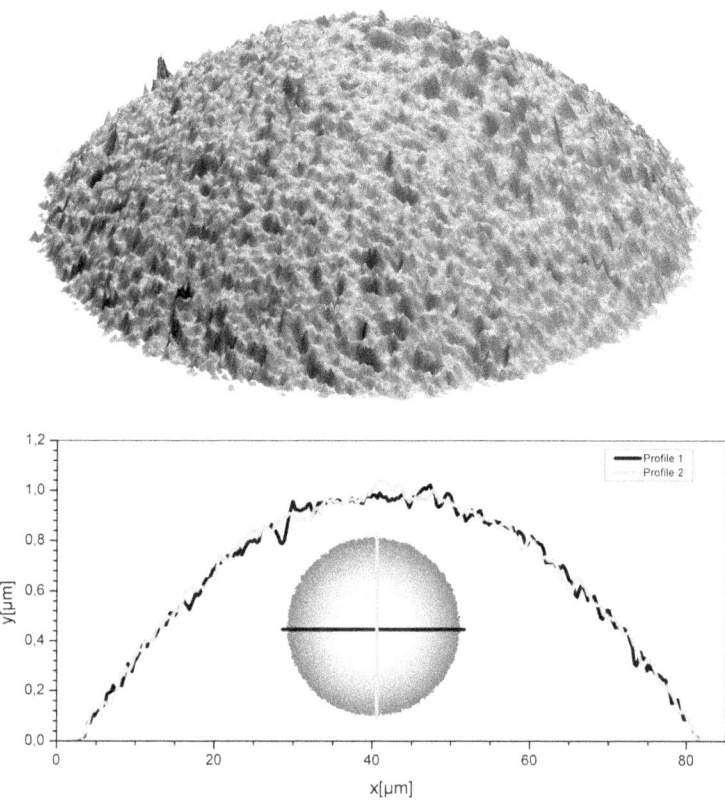

Figure A.4.: Li-HOSO in material combination sapphire ball on steel disc—spacial view created with withe light interferometry in the upper part and profiles in the lower part with profile sections as exhibited in the miniature

B. Additional information pertaining section 5.4.3

Table B.1.: Preset values of shear stress in frequency sweeps

Grease label	shear stress $\tau[\text{Pa}]$
HDS PAO	10.0
HDS HOSO	40.0
HDS OCT	10.0
HDS TMPO	10.0
BT PAO	20.0
BT HOSO	10.0
BT OCT	30.0
BT TMPO	15.0
Ca PAO	10.0
Ca HOSO	15.0
Ca OCT	15.0
Ca TMPO	15.0
Li PAO	10.0
Li HOSO	10.0
Li OCT	10.0
Li TMPO	40.0

C. Scientific publications in this work

C.1. Journal articles derived from this work

The following journal publications resulted from the scientific research of this work with their content presented in chapter 5:

1. Fiedler, M., Kuhn, E., Franco, J., Litters, T.: Tribological properties of greases based on biogenic base oils and traditional thickeners in sapphire-steel contact. Tribology Letters 44 (2011) p. 293-304

2. Fiedler, M., Sanchez, R., Kuhn, E., Franco, J.: Influence of oil polarity and material combination on the tribological response of greases formulated with biodegradable oils and bentonite and highly-dispersed silica acid. Lubrication Science 25 (2013) p. 397-412

C.2. Congress communications derived from this work

1. Fiedler, M.: Tribological Potential of Greases based on biogenous Esters. Tribology and Design, Faro Portugal, 11.-13.05.2010

2. Fiedler, M.: Tribologische Charakterisierung biogener Schmierfette. Tribologie Meeting 2010 - NMI Workshop, Haigerloch 16.-17.06.2010

3. Fiedler, M.: Tribological Potential of Greases based on biogenous Esters. 6th Arnold-Tross-Kolloquium, Hamburg 2010

4. Fiedler, M.: Tribologisches potential von Schmierfetten auf Basis biogener Grundöle. GfT Tribologie-Fachtagung 2010, Göttingen 27.-29.09.2010

5. Fiedler, M.: Bentonit und hochdisperse Kieselsäure als Verdicker für Biofette. 7th Arnold-Tross-Kolloquium, Hamburg 2011

6. Fiedler, M. Giese, C.: Kieselgele und Bentonite als biologisch inerte Verdickertypen von Schmierfetten auf Basis biogener Grundöle. GfT Tribologie-Fachtagung 2011, Göttingen 26.-28.09.2011

ORIGINAL PAPER

Tribological Properties of Greases Based on Biogenic Base Oils and Traditional Thickeners in Sapphire-Steel Contact

Martin Fiedler · Erik Kuhn · José María Franco · Thomas Litters

Received: 5 April 2011 / Accepted: 18 August 2011 / Published online: 28 August 2011
© Springer Science+Business Media, LLC 2011

Abstract Friction and wear tests were performed with a number of greases based on biogenic esters and thickened with two metal soaps and a highly dispersed silica acid gel. The series of experiments was performed on a Nonotribometer in material combination of sapphire ball on steel disks with a range of normal loads from 1 up to 500 mN. Results directly show influences of the bulk grease components on frictional and wear behavior. Comparison of frictional and wear results makes manifest that, while in most combinations of base oil and thickener, the highest influence is found in the thickening agent, some combinations are mainly influenced by the base oil. All frictional results along with wear widths and depths as well as micrographs of the prevailing wear mechanisms are presented and discussed.

Keywords Biodegradable oils · Organic esters · Friction mechanisms · Wear mechanisms · Abrasive wear · Adhesive wear

M. Fiedler (✉) · E. Kuhn
Laboratory of Machine Elements and Tribology, Department of Mechanical Engineering and Production, Hamburg University of Applied Sciences, Berliner Tor 5, 20099 Hamburg, Germany
e-mail: martin.fiedler@haw-hamburg.de

J. M. Franco
Complex Fluid Engineering Laboratory, Departamento Ingeniería Química, Facultad de Ciencias Experimentales, Universidad de Huelva, Avenida de las Fuerzas Armadas, S/N., 21071 Huelva, Spain

T. Litters
Fuchs Europe Schmierstoffe GmbH, Friesenheimer Str. 19, 68169 Mannheim, Germany

1 Introduction

Although undiscovered fossil hydrocarbon occurrences are predicted and technological progress will for some part increase the degree of feasibility of these resources, we will inevitably face the limits of availability of petrochemical resources in future [1, 2]. This will require a large change of ideas in the production of auxiliary and process materials in a large number of engineering fields. Consequently, a demand is made to focus on the use of renewable materials. Moreover, the steadily growing sense of ecological responsibility in society as well as legislation in concern of dealing with substances, which may have lasting harmful effects on environment puts a challenge to producers and users of lubricants to fall back upon biologically compatible alternatives.

It is stressed that lubricants are environmentally highly pertinent substances especially with regard to atmospheric carbon dioxide balance [3]. Also it is commonly known that vegetable oils more readily biodegrade. It was found that owing to their lower molecular weight, vegetable-derived-lubricants are significantly more degradable by maritime microbial communities than their mineral-derived-lubricant counterparts [4]. In combination of these demands, one can easily detect a trend toward regrowing biodegradable process materials.

Lubricating greases are produced as complex multiphase systems in which all component properties are needed for fulfillment of the desired function. In order to comply with all desired functions base oil as well as a thickening agent are needed, the latter of which may take a share of up to 50% and more depending on the required consistency and all the entailed characteristics of the grease. The presence of a metallic soap as a thickener does not only define the consistency of the grease but it also influences its frictional

 Springer

properties [5–9]. When formulating greases, which fully meet both above-mentioned requirements of biogenity and biodegradability, the manufacturer is challenged to take all of this into consideration for the selection of all the ingredients. If, for reasons of feasibility and lack of experience, not all components may be replaced by such of biogenic origin the utilization of biogenic base oil may at least be a step into the right direction. In the present article, some greases based on biogenic base oils are tribologically characterized.

There are only a few studies found in the literature, which directly examine the influences of main components of biodegradable greases on frictional and wear behavior. In this study, greases formulated with combinations of three different biogenic base oils and classical thickener types are compared with each other and to a grease system based on synthetically generated polyalphaolefin (PAO) thickened with the same thickening agents. Results appear to elucidate the role of both the employed base oils and thickener agents in frictional and wear behavior.

2 Experimental Procedures

2.1 The Greases

Three different biogenic and biodegradable base oils mainly consisting of organic esters, high oleic sunflower oil (HOSO), octyldodecyl isostearate (OCT), trimethylolpropane trioleate (TMPO), and one synthetic biodegradable low viscous PAO as reference oil with basic characteristics displayed in Table 1, were used as base oils to formulate the lubricating greases studied.

Lithium and calcium soaps both of which are formulated as hydroxy stearates, as well as a highly dispersed silica acid (HDS) gel thickener were used as thickener agents. No additives were applied in the formulation except for swelling aid in the case of HDS greases. Composition and characteristics of the greases are depicted in Table 2.

2.2 The Testing Equipment

Tribological friction and wear tests for all greases were performed on a CSM nanotribometer as shown in Fig. 1. This ball-on-disk apparatus works in linear oscillation and circular rotation mode. In this study, all tests were performed in the rotational mode. The velocity of relative motion between the static ball and the dynamic disk directly depends on the diameter of the set wear track in correlation to the set rpm resulting in gliding friction. The

Table 1 Base oil characteristics

	v_{40} [mm^2 s^{-1}]	v_{100} [mm^2 s^{-1}]	Density [g/cm^3]	Degradability[a]
HOSO	38.8	8.5	0.92	96%
TMPO	48.0	9.8	0.92	96%
OCT	25.5	5.5	0.87	97%
PAO (synth. reference oil)	46.7	7.9	0.83	Yes[b]

[a] According to OECD 301B

[b] According to Materials Safety Data Sheet it is expected to be inherently biodegradable

Table 2 Grease characteristics

Grease label	Thickener		Pu[a] [1/10 mm]	Pw[a] [1/10 mm]	NLGI class	DP[b] [°C]	Oil separation[c] [wt%]
	Type	Content [%]					
HDS PAO	Highly dispersed silica acid	10.3	269	280	2	–	1.12
HDS HOSO		13.9	238	268	2	–	0.92
HDS OCT		11.4	264	287	2	–	1.22
HDS TMPO		14.1	249	287	2	–	0.82
Ca PAO	Ca-stearate	18.8	283	287	2	133	1.16
Ca HOSO		17.6	272	279	2	132	0.80
Ca OCT		15.7	253	272	2	136	0.88
Ca TMPO		13.7	260	272	2	132	0.63
Li PAO	Li-12-hydroxy stearate	9.7	279	283	2	205	2.57
Li HOSO		15.9	264	272	2	195	1.12
Li OCT		13.6	279	294	2	188	1.79
Li TMPO		20.7	264	279	2	190	0.81

[a] Unworked and worked penetration according to DIN ISO 2137

[b] Dropping point according to DIN ISO 2176

[c] According to DIN 51817

2.3 Tribological Tests

All of the tribological examinations presented below were performed on the above-described nanotribometer in rotational mode with a relative gliding speed of 5.0 mm s^{-1}, using a sapphire ball with 1.5-mm diameter and steel plates (115CrV3, hardness 22.72 HRC). All the steel plates were metallographically grinded and polished with a soft finish diamond paste of particle size 3 μm in the last polishing step. The nanotribometer worked in ambient pressure and temperature. In the test procedures, the normal force was varied with resulting Hertzian stress values as shown in Table 3.

At first, a grease layer of 0.05 mm was applied to the steel plate then the measurement parameters of normal force, relative speed, and track radius were set up. The test duration was 50 min each. In order to allow conclusions to be drawn about the friction and wear influence of each bulk component of the greases the above-described tests were performed with all the greases, each base oil on its own and in completely dry contact situation.

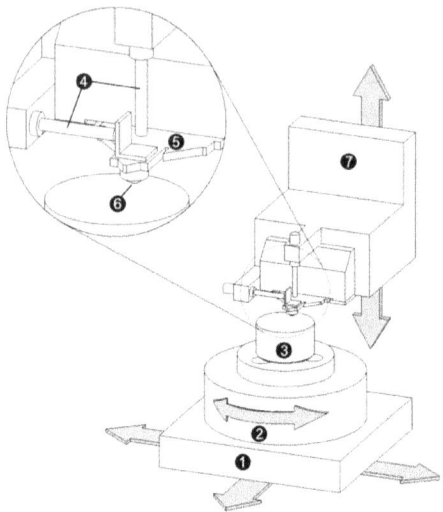

Fig. 1 Experimental set-up—nanotribometer Note: *1* XY-cross-table, *2* Rotational axis, *3* Specimen steel disk, *4* Interferometrical sensors, *5* Cantilever, *6* Specimen sapphire ball, *7* Z-step-motor-axis with integrated piezo-axis

step motor of the rotational axis makes up to 120 rpm and is mounted to an XY-cross-table, which positions the disk under the ball to set up a controlled track radius. Specimen balls are fixed to a cantilever module and normally positioned to the plate by a Z-step-motor-axis. Defined load is applied to the ball by fine positioning and deflecting the cantilever via piezo-axis. The deflection of the cantilever is interferometrically measured and then controlled. By means of two distally attached mirrors, the deflection of the cantilever in radial (frictional force) as well as normal (normal force) direction of the plate revolution is detected and by the very precise calibration of the spring constant of the cantilever in both main directions calculated into frictional and normal forces. Both of these forces, the coefficient of friction, the relative indention of the ball into the plate, as well as the ambient temperature and the relative humidity of the measuring cabin are detected and recorded during the measurements at a set sample rate. The utilization of cantilevers with different spring constants covers a large load range from 0.1 up to 500 mN. The high-precision optical interferometric measurement detects the wear height with a resolution of 50 nm. Despite relatively low normal forces of the nanotribometer, high values of Hertzian stress (up to ~2GPa) can be applied to the measured tribological system. This is due to the fact that very small balls may be applied to the cantilever.

3 Results

3.1 Friction

In general, friction coefficients vary with high fluctuation in most frictional tests. Averaged coefficients of friction of all the examined greases with material combination sapphire ball on steel plate are shown in Fig. 2. Frictional results in the present study have an average standard deviation of more than 10%. Fluctuations in values of coefficients of friction are lower for higher normal forces of 500 and 100 mN. Values for 10 mN and less in fact are so unstable that they become useless. For this reason, the depiction of frictional results for normal forces of 10 and 1 mN is renounced. This typical effect of the nanotribometer is caused by rheological influences evoked by grease layer displacement at every revolution of the steel plate. For this reason, it is very important to always apply grease layers of the same thickness to the steel disks. Thicker grease layers evidently cause higher resisting abilities toward repositioning by the specimen ball.

In attempting to attribute frictional values to bulk grease components, one has to regard results from each component's point of view. At first consideration from the

Table 3 Test conditions

Normal force [mN]	500	100	10	1
Max Hertzian stress [GPa]	1.52	0.89	0.41	0.19

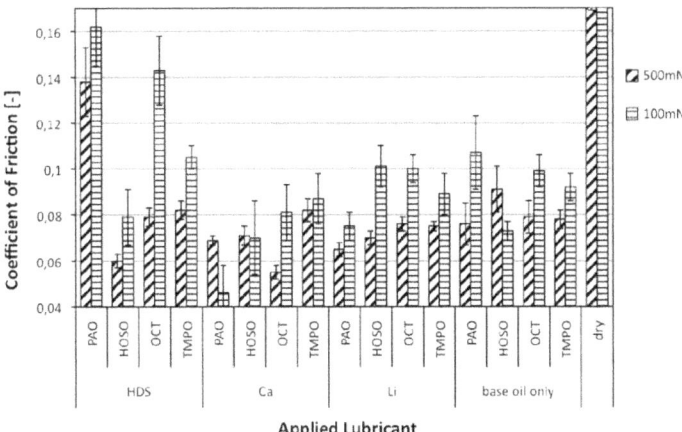

Fig. 2 Friction in contact situation sapphire ball on steel disk

thickener's point of view, one detects a general tendency of the highest frictional values for the series of HDS-thickened greases. The group of calcium-thickened greases stands out because of its lowest friction coefficients in comparison, followed by lithium-thickened greases.

When comparing component attributes and their influence on friction from the base oil point of view, one first of all finds out that the ranking within the group of base oils from PAO with highest frictional values down to HOSO with the lowest coefficients of friction is also found in the group of greases formulated with HDS-gel-thickener with this tendency even more intensified. Reasons for this effect will be discussed later on with respect to wear intensities and mechanisms. It also shows that the group of base oils presents higher frictional values than Ca-soap-thickened formulations in the present frictional contact situation sapphire ball on steel plate. Moreover, the ranking that was found for base oils is almost completely turned to the opposite tendency now with least coefficients of friction for Ca-PAO. Results also show that influences of single base oils on Li-thickened greases are too small to be detected or even interpreted into tendencies. The only conspicuousness is found for PAO thickened with Li-soap—here again it results in smallest coefficients of friction.

3.2 Wear

Different behaviors of grease formulations and their components are generally more explicitly displayed in wear than in frictional results. In the same way, wear measurements fluctuate less than frictional measurements. Considering the full dimensions of wear only after the completion of a frictional measurement implies to break a process factor down into a static view. This way fluctuation is eliminated to some extent. However, it is also important to keep in mind that friction and wear always depend on each other in the total frictional system. In this series of experiments, wear was analyzed statically and process orientated. For the static part, which was measured after the tests, the widths of all the wear tracks on steel disks and the corresponding diameters of all the wear scars on sapphire balls are displayed for all the measured normal forces in Fig. 3. This diagram shows that influences of bulk components differ with the combination of materials in the tribological system. Micrographs of all the wear marks produced with normal load conditions from 500 down to 10 mN are displayed in Figs. 5, 6, and 7. As it has been done previously with the interpretation of frictional results, wear characteristics are also associated with all single bulk grease components. Sorting wear characteristics according to thickening agents shows that ball and disk wear need to be evaluated separately. Some greases e.g., produce high rates in ball wear but leave the disks unaffected and vice versa. It also becomes apparent that each normal force has its own characteristic influence on wear behavior so it does not always prove to be true when simply predicting less wear as a result of less normal force.

The most evident fact for HDS-based greases is that they produce the widest wear scars in sapphire balls, as they are even wider than in completely dry contact situation. For a normal load of 500 mN, this same effect takes place on the surface of the steel plates, resulting in the widest wear tracks measured in the whole series of experiments ranging from 35 μm in case of HDS–OCT up to 62 μm for HDS–PAO. The smaller the normal loads, however, the more these tendencies diverge. While wear scars on the sapphire

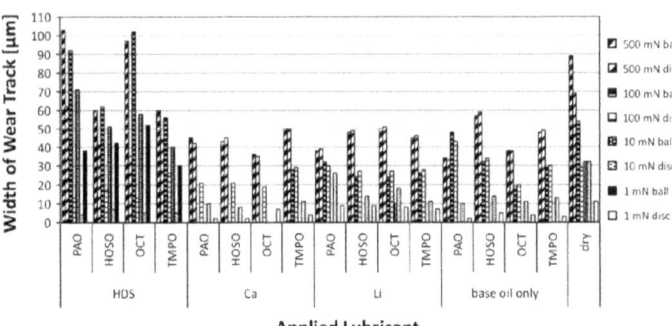

Fig. 3 Width of wear tracks—sapphire ball on steel disk

balls remain the biggest compared to all the other lubricants measured and even when compared to dry contact, the wear tracks on the steel plates become smaller and smaller. This effect becomes the most evident for 1 mN, the smallest of all normal forces applied in this series of experiments. At this load HDS-thickened greases still produce very wide wear scars ranging from 30 μm in case of HDS–TMPO to 52 μm in diameter for HDS–OCT (displayed in black solid bars in the diagram in Fig. 3), while, on the other hand, they do not produce any detectable wear on the steel plates. For soap-thickened greases and base oils only, there is an opposite effect resulting in no scars on the balls with still evident and measurable wear tracks on the plates (white solid bars in the diagram). Moreover, less normal force in HDS-based greases results in much less wear compared with soap greases—wear-force gradients in HDS-thickened greases seem to differ from those in metal soap thickened ones. With all the metal soap greases and base oils only, measured at 500 mN normal force, ball wear equals disk wear with similar wear widths ranging from 33 to 59 μm. Moreover, at these lubricants, with the exception of Li-OCT, there is no detectable wear in sapphire balls at 10 mN normal force. It also becomes clear that at normal forces of 100 mN and less Calcium greases produce the least if any wear in sapphire balls. The only Ca-grease formulation that occasions wear in the sapphire balls is Ca-TMPO. The same thing applies to disk wear at a normal force of 100 mN, here also calcium greases produce the lowest values ranging from 19 μm in case of Ca-OCT to 29 μm for Ca-TMPO. At normal forces of 10 mN, Calcium greases together with base oils only produce least wear in disks with values ranging from 8 μm in case of Ca-HOSO to 14 μm for HOSO base oil only. The test results for Li-greases show that for normal loads of 500 and 100 mN, each width of ball wear approximates to the corresponding wear in the disks with values extending from 38 to 51 μm in 500 mN tests and from 24 to 32 μm at 100 mN of normal load. On the other hand, for 10 mN and less, the only wear to be detected is located on the disk. The only exception to this rule is found for Li-OCT. Similar behavior is observed with Ca-TMPO. The outcomes of tests with applied loads of 500 and 100 mN on base oils only reveal that, just like within the group of Li-greases, the widths of ball wear resemble the according dimensions of disk wear with values stretching from 33 to 59 μm at 500 mN loads and from 19 to 48 μm at normal forces of 100 mN. It also shows that with applied loads of 10 mN in the group of base oils, there is only disk wear revealing values of the same sequence as with higher loads, which is HOSO, TMPO, PAO, and OCT from the highest to the lowest. In the upper spectrum with normal loads of 500 and 100 mN in dry contact situation with sapphire balls on steel disks, wear widths decrease with lower normal loads. At these loads, wear tracks on the disks are 22–46% narrower than those on the corresponding balls. In the lower spectrum of normal loads of 10 and 1 mN in unlubricated contact, however, widths of wear tracks do not decrease with smaller forces in the case of sapphire balls. In case of 10 mN, wears in both contact partners are in equilibrium.

When now sorting wear characteristics according to the applied base oils in the formulation, one soon finds out that the influence of the base oil on wear is much smaller than that of the thickener. It also becomes apparent that the base oil influences differ according to the formulated thickener. Thus, it shows that within the group of HDS-greases, HOSO and TMPO oils generate the least wear, while PAO and OCT originate the highest wear widths. On the other hand, in the tests of plain base oils, there are opposite tendencies to be detected with HOSO and TMPO producing the highest wear widths.

The above-mentioned process-orientated analysis of wear experiments was performed as in situ measurement in every single frictional test on the nanotribometer. As described earlier, the nanotribometer monitors wear in the form of indention of the specimen ball into the testing

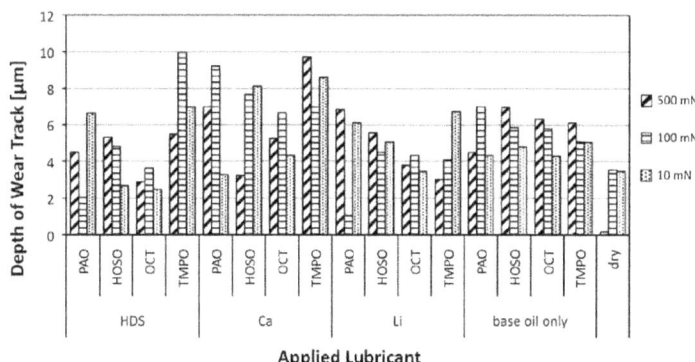

Fig. 4 Depth of wear tracks—sapphire ball on steel disk

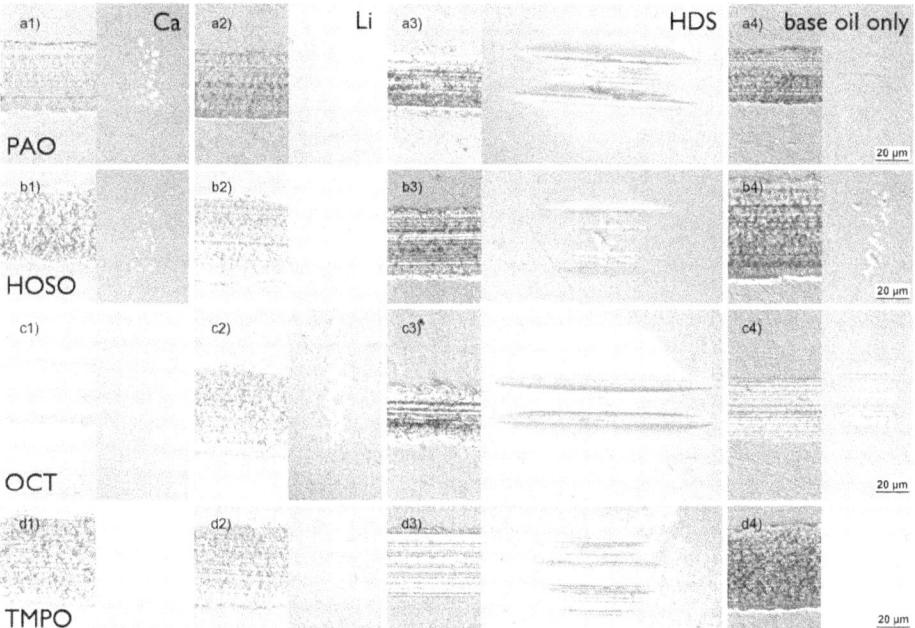

Fig. 5 Matching sets of disk and ball wear—normal load 500 mN, sapphire balls on steel disks Note: Grease compositions and base oils only—sorted by applied thickener in columns and base oil in lines

surface. This way it measures the total depth of penetration but does not, however, judge these indention results according to the location of wear in the surfaces. In other words these wear penetration results do not give any information about whether the wear took place in the ball, the disk or intermediate layers. Measurement results show a steadily declining course of values as expected for indention. Because of the regularity of the decrease of values only the final values are displayed in this study (see Fig. 4).

The biggest noticeable difference between the depths and the widths of wear tracks in this series of experiments

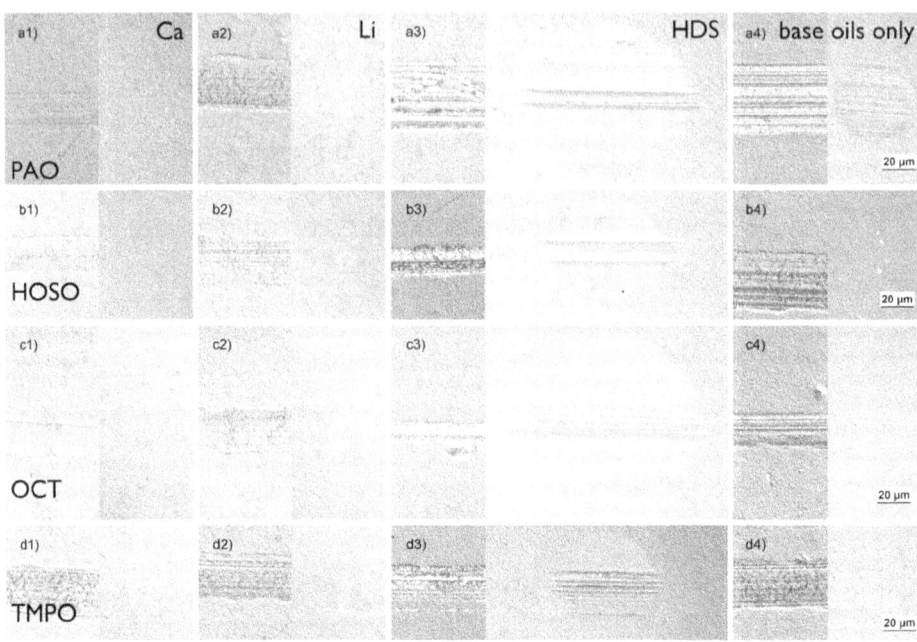

Fig. 6 Matching sets of disk and ball wear—normal load 100 mN, sapphire balls on steel disks Note: Grease compositions and base oils only—sorted by applied thickener in columns and base oil in lines

is the relatively low penetration depth of HDS-greases. In the measurements of wear widths, the values for HDS-greases jutted out of all other results, while in the depth of wear, they take the lowest segment. The only exception to this rule is HDS–TMPO, which is ranked with the highest values. Another conspicuousness about theses results is the fact that categorization by thickening agents, which showed distinct differences between the single groups in the analysis of wear widths does not induce the same effect when evaluating the wear depths. What also becomes clear when considering wear depth results categorized by thickener is that the group of Ca-greases now reveals the highest wear depths, while during the analysis of wear widths, it had the lowest rates. Moreover, in contrast to the evaluation of wear widths, now base oil results show quite steady behavior and do not seem to give any room for interpretation toward a ranking. Probably, the most peculiar and unexpected outcome of this analysis are the very low indention depths measured for the completely dry contact. Categorizing the indention depths by the formulated base oil results in a ranking that starts with the highest values for TMPO (6.48 ± 2.14 µm), followed by HOSO (5.38 ± 1.62 µm), PAO (5.21 ± 2.36 µm), and OCT (4.44 ± 1.33 µm).

4 Discussion

At the examination of friction and wear as two interrelating dimensions, it is insufficient only to take frictional values and wear widths and depths into consideration to receive significant information about the tribological system. When trying to understand and correctly evaluate friction and wear, one must also comprehend their states, types, and especially their mechanisms. Different wear states, as defined by Fleischer [10] and Czichos [11], evoked by specific wear types are made manifest through different wear mechanisms, which make a deep impact on frictional and wear behavior. As a consequence, different wear mechanisms result not only in different wear heights and widths but also in altering frictional values. Moreover, it is possible even to have several wear mechanisms present in one tribological contact system. Micrographs of wear scars on the disks and balls of all tribological systems examined

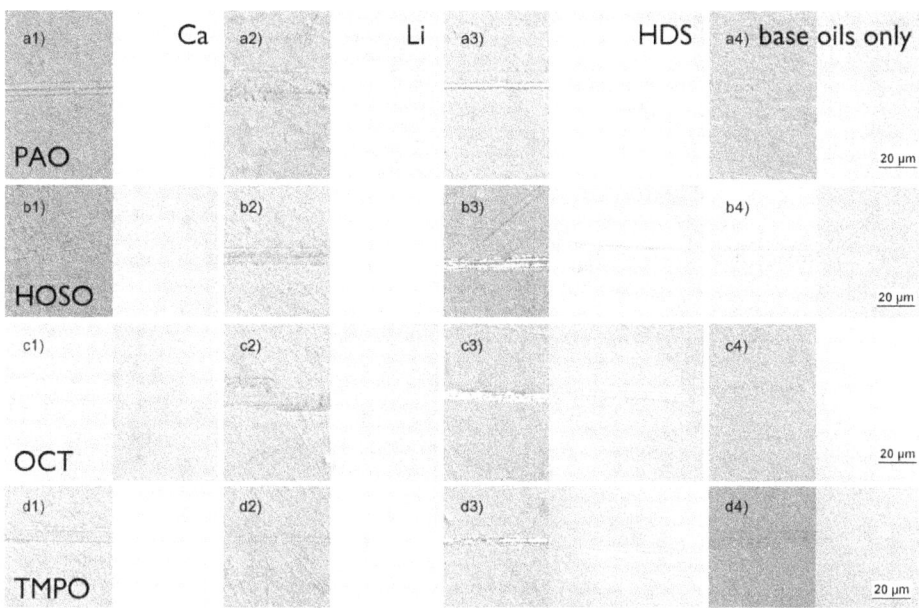

Fig. 7 Matching sets of disk and ball wear—normal load 10 mN, sapphire balls on steel disks Note: Grease compositions and base oils only—sorted by applied thickener in columns and base oil in lines

earlier in this study are displayed in Figs. 5, 6, and 7; they supply an opportunity to interpret prevailing wear mechanisms. Wear widths and depths depicted in these micrographs were analyzed earlier in this study and are only to be mentioned again with reference to comparison of images and data presented earlier in diagrams. When evaluating these images according to wear mechanisms, one has to be very careful to only objectively consider present facts. These facts are e.g., the number and the intensity of wear grooves when appraising abrasive processes, number, and size of fretting particles that adhere to counter surfaces when assessing the adhesive wear process and the number with its accompanying size of cracks and their intensity orthogonal to the direction of relative motion which point to the fact that a surface delamination process took place in the tribological system. Measuring depth profiles of the wear tracks would help in analyzing the only just mentioned intensities. However, profile measurements were not possible in this study, and so intensities of wear mechanisms will be rated and set in relation to each other as displayed in Tables 4, 5, and 6. Every single figure has its own rating so that intensities are compared to the average of each single figure with a rating from less than average up to superior impact (see notes of Tables 4, 5, 6).

Summing up the image interpretation of all wear marks in Figs. 5, 6, and 7, one may detect a good correlation to the frictional and wear results presented earlier in this study. Within the group of Ca-greases, the images reveal surface delamination as well as plastic deformation and abrasive wear mechanisms in the disks and mostly exclusively adhesion mechanisms in the sapphire balls. The same mechanisms occur in the group of Li-greases with generally higher intensity. This behavior can be elucidated only when taking the microphysical mode of action of lubrication into consideration. Lubrication only comes into effect by the application of a lubricant film that mechanically separates the contact partners [9, 11]. Both, base oil and thickening agent influence the formation of this separating film in lubricating grease [7, 12]. The results of performed series of experiments show that the thickener's influence is more substantial in this context. To some extent, explanations for these effects can be found in the microstructure of the thickening agents. Metallic soaps usually form three-dimensional networks [13] of polymorphic soap fibers that confine the oil with secondary valence bonds. It is well known that lithium-greases generate longer and more densely arranged and entangled fibers than calcium-greases [14–17]. This difference could be an explanation for the

Table 4 Wear mechanisms in sets of disk and ball wear—normal load 500 mN, sapphire balls on steel disks

	Ca	Li	HDS	Base oil only
PAO	Disk	Disk	Disk	Disk
	Surface delamination	Surface delamination+	Plastic deformation++	Surface delamination
	Plastic deformation	Plastic deformation	Abrasion+++	Plastic deformation
	Abrasion+	Abrasion+		Abrasion
	Ball	Ball	Ball	Ball
	Adhesion+	Adhesion−	Adhesion+	Abrasion+
			Abrasion+++	
HOSO	Disk	Disk	Disk	Disk
	Surface delamination+	Surface delamination+	Plastic deformation+	Surface delamination+
	Plastic deformation+	Plastic deformation+	Abrasion+++	Plastic deformation+
	Abrasion−	Abrasion+		Abrasion
	Ball	Ball	Ball	Ball
	Adhesion	Adhesion+	Adhesion+	Adhesion+
	Abrasion+	Abrasion+	Abrasion+	Abrasion+
OCT	Disk	Disk	Disk	Disk
	Plastic deformation−	Surface delamination+	Plastic deformation++	Abrasion++
	Abrasion−	Plastic deformation+	Abrasion+++	
		Abrasion+		
	Ball	Ball	Ball	Ball
	No wear marks	Adhesion+	Adhesion+	Adhesion+
			Abrasion+++	Abrasion++
TMPO	Disk	Disk	Disk	Disk
	Surface delamination+	Surface delamination+	Plastic deformation+	Surface delamination+
	Plastic deformation+	Plastic deformation+	Abrasion++	Plastic deformation+
	Abrasion−	Abrasion+		Abrasion
	Ball	Ball	Ball	Ball
	Adhesion+	Adhesion+	Abrasion+	Adhesion−
				Abrasion+

Notes: Relative rating in comparison to each other: − less than average; + more than average; ++ much more than average; +++ superior impact

lower coefficients of friction and the generally lower wear rates of examined calcium-greases compared to the greases formulated with lithium-soap. (Table 5).

The group of HDS-gel-greases tested under all normal load conditions presents itself with much different wear marks than those known for soap greases. In all sapphire balls used to measure single greases, there are marks of the highest abrasive wear in the central region surrounded by a circular area of homogenous fine particle erosion that almost looks like polished material. The surface roughness of these polished areas, even though not measured, stands out from the processed surface of the rest of each sapphire ball. The central grooves of the strongest abrasive wear in the sapphire balls created deep highly abrasive marks in the corresponding surfaces of the steel disks. The fact that apart from these deep marks, the steel surfaces appear completely unspoiled, while on the surfaces of the sapphire counter parts, there are much larger worn-off areas is the most conspicuous for the group of HDS-greases. Again, reasons for this effect have to be found in the microstructure of the thickener. HDS-thickener mostly consists of SiO_2 particles of size 0.1 up to 1.0 μm [18]. The presence of theses particles has been proven in the residue of wear debris via EDX-analysis [16, 17]. The high proportion of abrasive wear with the use of HDS-greases especially in the sapphire balls is to be explained by the extreme hardness of SiO_2 particles (5–7 on the hardness scale according to Mohs). Although sapphire is even harder (up to 9 on the Mohs scale), it yields the persistent influence of SiO_2 particles. It is assumed that these SiO_2 particles are elastically embedded in the metal surface of the steel plates during the course of performed frictional tests without abrasively influencing the steel. This would explain why, in some parts, the sapphire balls wore off, while the corresponding contacted area in the steel plates was left unaffected.

Table 5 Wear mechanisms in sets of disk and ball wear—normal load 100 mN, sapphire balls on steel disks

	Ca	Li	HDS	Base oil only
PAO	Disk	Disk	Disk	Disk
	Plastic deformation	Surface delamination+	Plastic deformation+	Abrasion++
	Abrasion	Plastic deformation+	Abrasion+++	
		Abrasion+		
	Ball	Ball	Ball	Ball
	No wear marks	Adhesion+	Abrasion+++	Abrasion++
HOSO	Disk	Disk	Disk	Disk
	Surface delamination−	Surface delamination	Plastic deformation+	Surface delamination+
	Plastic deformation	Plastic deformation	Abrasion++	Plastic deformation+
	Abrasion−	Abrasion		Abrasion+
	Ball	Ball	Ball	Ball
	No wear marks	Adhesion	Abrasion++	Adhesion+
				Abrasion
OCT	Disk	Disk	Disk	Disk
	Plastic deformation−	Surface delamination	Plastic deformation+	Surface delamination
	Abrasion−	Plastic deformation	Abrasion+	Plastic deformation
		Abrasion		Abrasion
	Ball	Ball	Ball	Ball
	No wear marks	Adhesion	Abrasion+++	Abrasion+
TMPO	Disk	Disk	Disk	Disk
	Surface delamination+	Surface delamination+	Plastic deformation+	Surface delamination+
	Plastic deformation+	Plastic deformation+	Abrasion+++	Plastic deformation+
	Abrasion	Abrasion+		Abrasion+
	Ball	Ball	Ball	Ball
	Adhesion+	Adhesion	Abrasion++	Adhesion−
				Abrasion+

Notes: relative rating in comparison to each other: − less than average; + more than average; ++ much more than average; +++ superior impact

Because of major differences in the prevailing wear mechanisms in the group of base oils as sole lubricant, there seems to be smaller correlation between wear and friction. There is slight abrasive wear in the ball and the disk and surface delamination in the disk evoked by the use of PAO. The same mechanisms in addition to adhesive wear work with HOSO. OCT is the only base oil that exclusively causes abrasive wear on both frictional partners. TMPO on the other hand only produces surface delaminations in the steel plate with only marginal abrasive and adhesive wear in the ball.

Although in the formulated greases the general influence of base oils on friction and wear only played a subordinate role, there are still some observations worth mentioning. Since all tests were performed in rotational mode, there might be a tendency toward starvation of the frictional track. It is assumed that the lower the viscosity of the base oil, the more it tends to flow back into the frictional track. Comparing the base oil viscosities to frictional and wear results, however, substantiates this assumption only to the extent of a general tendency.

Wear results also make clear that base oils on their own behave differently in the tribocontact than base oils in formulation with different thickeners. The OCT ester and PAO measured in 500 and 100 mN of normal load regime are the oils with the highest rates of abrasive wear in the sapphire balls. This characteristic seems to be even more intensified in combination with HDS-gel-thickeners where they result in the widest wear diameters. In combination with Li-soap-thickener, however, this characteristic is limited to the same extent of all other base oils; in addition, most interestingly, Ca-thickener seems to change this effect to the opposite in the case of OCT-ester resulting in the smallest wear diameters with the least abrasive wear. These results directly correlate to the frictional examinations displayed in Fig. 2. Microstructural analysis of some exemplary Li-greases by means of atomic force microscopy may help elucidate these effects. AFM images displayed in Fig. 8 show a direct influence of the base oil on

Table 6 Wear mechanisms in sets of disk and ball wear—normal load 10 mN, sapphire balls on steel disks

	Ca	Li	HDS	Base oil only
PAO	Disk	Disk	Disk	Disk
	Abrasion+	Surface delamination+	Plastic deformation+	Plastic deformation+
		Plastic deformation	Abrasion++	Abrasion
		Abrasion		
	Ball	Ball	Ball	Ball
	No wear marks	No wear marks	Abrasion+++	No wear marks
HOSO	Disk	Disk	Disk	Disk
	Abrasion	Surface delamination	Plastic deformation+	Surface delamination
		Plastic deformation	Abrasion+++	Plastic deformation
		Abrasion		Abrasion++
	Ball	Ball	Ball	Ball
	No wear marks	No wear marks	Abrasion ++	Adhesion
				Abrasion
OCT	Disk	Disk	Disk	Disk
	Abrasion	Surface delamination	Plastic deformation+	Abrasion
		Plastic deformation+	Abrasion++	
		Abrasion		
	Ball	Ball	Ball	Ball
	No wear marks	Adhesion	Abrasion+++	Adhesion+
TMPO	Disk	Disk	Disk	Disk
	Surface delamination−	Abrasion	Plastic deformation+	Surface delamination+
	Plastic deformation−		Abrasion++	Abrasion+
	Abrasion+			
	Ball	Ball	Ball	Ball
	No wear marks	No wear marks	Abrasion++	Abrasion

Notes: relative rating in comparison to each other: − less than average; + more than average; ++ much more than average; +++ superior impact

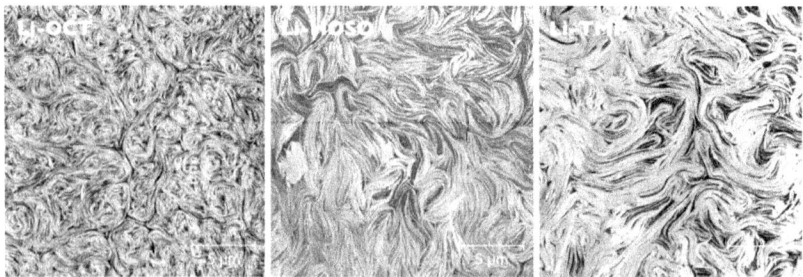

Fig. 8 AFM—micrographs of some exemplary Li-greases

the formulation of soap fibers. In combination with HOSO and TMP, the Li-soap resulted in long, mostly parallel and less entangled fibers. However, in combination with OCT, it resulted in rather short and more entangled fibers. As explained earlier in this study, it is assumed that the shorter soap fibers result in less friction and harmful wear mechanisms. In deduction, it is assumed that the less

abrasive character of shorter soap fibers in general and specifically in combination of Li-soap-thickener and OCT base oil counteracts the higher abrasive properties of OCT resulting in less-to-equal wear intensities and frictional values.

Different behaviors between HDS greases and base oils only provide some examples for diverging tendencies.

With all HDS-greases, wide wear tracks result in high values of coefficients of friction—with base oils only, this effect is turned to the opposite—and the high values in wear width result in low friction.

5 Conclusions

While highly dispersed silica acid gel thickener most substantially affected the wear behavior toward a highly abrasive nature, lithium and calcium thickeners mostly resulted in a combination of plastic deformation with slight surface delamination in the steel plates and adhesive wear in the sapphire balls. While base oil influences retain a high impact with the use of HDS-gel and Ca-soap as thickeners, they are limited in the group of greases thickened with of Li-soap. Synthetic PAO and biogenic OCT ester as base oils maximized the abrasive nature of HDS-gel thickener. OCT ester, on the other hand, minimized the effects of adhesion of particles to the sapphire ball surface and the intensity of surface delamination of the steel surface in combination with Ca-soap thickener. Ca-thickened grease formulations generally resulted in least wear and coefficients of friction. Lithium greases resulted in values for the coefficient of friction and wear widths similar to base oils, although wear depths and wear mechanisms were much different. HDS greases resulted in the highest wear rates and frictional values as well as the most destructive wear mechanisms.

Acknowledgments The research presented in this article was performed as a joint venture of the above-mentioned institutions as part of the TRIBIOS project, which is funded by the Federal Ministry of Education and Research Germany. The authors gratefully acknowledge the financial support.

References

1. Bentley, R.W.: Global oil and gas depletion: an overview. Energy Policy **30**, 189–205 (2002)
2. Rogner, H.H.: An assessment of world hydrocarbon resources. Annu. Rev. Energy Environ. **22**, 217–262 (1997)
3. Willing, A.: Lubricants based on renewable resources—an environmentally compatible alternative to mineral oil products. Chemosphere **43**, 89–98 (2001)
4. Mercurio, P., Burns, K., Negri, A.: Testing the ecotoxicology of vegetable versus mineral based lubricating oils: 1. Degradation rates using tropical marine microbes. Environ. Pollut. **129**, 165–173 (2004)
5. Yokouchi, A., Yamamoto, Y.: Influence of soap fiber structure on frictional property of lithium soap grease. ASME Conf. Proc. **2007**, 129–131 (2007)
6. Rong-Hua, J.: Effects of the composition and fibrous texture of lithium soap grease on wear and friction. Tribol. Int. **18**, 121–124 (1985)
7. Hurley, S., Cann, P.: Grease composition and film thickness in rolling contacts. NLGI Spokesm. **63**, 12–22 (1999)
8. Cousseau, T., Graça, B.M., Campos, A.V., Seabra, J.H.O.: Influence of grease formulation on thrust bearings power loss. J. Eng. Tribol. **224**, 935–946 (2010)
9. Kuhn, E.: Zur Tribologie der Schmierfette. expert Verlag, Renningen, Germany (2009). ISBN: 978-3-8169-2869-0
10. Fleischer, G., Gröger, H., Thum, H.: Verschleiß und Zuverlässigkeit. VEB Verl. Technik, Berlin (1980)
11. Czichos, H.: Tribologie-Handbuch. Vieweg und Teubner, Wiesbaden (2010)
12. Cann, P., Hurley, S.: Friction properties of grease in ehd lubrication. NLGI Spokesm. **66**, 6–15 (2002)
13. Delgado, M., Franco, J., Kuhn, E.: Effect of rheological behaviour of lithium greases on the friction process. Ind. Lubr. Tribol. **60**, 37–45 (2008)
14. Sánchez, M., Franco, J.M., Valencia, C., Gallegos, C., Urquiola, F., Urchegui, R.: Atomic force microscopy and thermo-rheological characterisation of lubricating greases. Tribol. Lett. **41**, 463–470 (2011)
15. Kimura, H., Imai, Y., Yamamoto, Y.: Study on fiber length control for ester-based lithium soap grease. Tribol. Trans. **44**, 405–410 (2001)
16. Fiedler, M., Käfer, E.: Tribologisches Potential von Schmierfetten auf Basis Biogener Grundöle. In: Gesellschaft für Tribologie e.V. (ed.) Reibung, Schmierung und Verschleiß – Forschung und praktische Anwendungen, pp. 57/1–57/12. Aachen (2010). ISBN: 978-3-00-032180-1
17. Fiedler, M.: Tribological potential of greases based on biogenous esters. In: Kuhn, E. (ed.) 6. Arnold Tross Kolloquium, pp. 229–269. Shaker Verlag, Aachen (2010). ISBN: 978-3-8322-9599-8
18. Goerz, T.: Gel-und Bentonitfette Zusammensetzung-Eigenschaften. Tribol. und Schmierungstechnik **56**, 30–34 (2009)

Influence of oil polarity and material combination on the tribological response of greases formulated with biodegradable oils and bentonite and highly dispersed silica acid

M. Fiedler[1,*,†], R. Sanchez[2], E. Kuhn[1] and J. M. Franco[2]

[1]*Laboratory of Machine Elements and Tribology, Department of Mechanical Engineering and Production, Hamburg University of Applied Sciences, Berliner Tor 5, 20099 Hamburg, Germany*
[2]*Complex Fluid Engineering Laboratory, Departamento Ingeniería Química, Facultad de Ciencias Experimentales, Universidad de Huelva, Huelva, Spain*

ABSTRACT

Different biodegradable lubricating greases formulated with esters of fatty acids, as base oils, and bentonite and highly dispersed silica acid, as thickener agents, were tribologically investigated in a nanotribometer and compared with polyalpha olefin greases with equal thickeners. Material combinations of steel ball on steel disc and sapphire ball on steel disc were used with different normal loads. Several friction and wear effects were found depending on the thickener and the base oil. The influence of grease components is also different in both material combinations evaluated. On the one hand, the base oil exerts a much higher impact on friction and wear in grease systems thickened with highly dispersed silica acid than in those thickened with bentonite. On the other hand, the latter reacts more sensitively to a change in material combination. Results were discussed and explained on the basis of polarity influences of the base oils and solid surfaces. Copyright © 2012 John Wiley & Sons, Ltd.

Received 25 May 2012; Revised 10 August 2012; Accepted 15 August 2012

KEY WORDS: biodegradable greases; base oil polarity; AFM; sliding wear; micro-scale abrasion

INTRODUCTION

In the background of a steadily growing awareness of ecological regards in legislation as well as society, a special focus should be drawn to the twofold ecological impact of lubricating greases based on renewable oleochemicals in many engineering fields. These lubricants may not only play a superior role in the reduction of energy losses caused by friction in tribosystems of many machine elements but also minimise wear. The second, and in the scope of this study even more important, ecological impact is based on their biological degradability with its resulting elimination of direct and indirect environmental pollution caused by a closed-loop carbon dioxide derivation cycle. Due to these reasons, it is reported that

---

*Correspondence to: M. Fiedler, Laboratory of Machine Elements and Tribology, Department of Mechanical Engineering and Production, Hamburg University of Applied Sciences, Berliner Tor 5, 20099 Hamburg, Germany.
†E-mail: martin.fiedler@haw-hamburg.de

Copyright © 2012 John Wiley & Sons, Ltd.

oleochemicals have a sustainable impact on global atmospheric carbon dioxide balance.[1] With respect to triblogical characteristics, vegetable base oils in many cases do not show any disadvantages to their synthetically or petrochemically derived counterparts.[2,3] Moreover, many of the main negative features encountered with the use of biogenic esters as base oils such as oxidative, ageing and thermal stability may be overcome to some degree by enzymatic or catalytic modification (transesterification, selective hydrogenation and oligomerisation) of the vegetable oil esters.[2] To produce bio-lubricating greases, the replacement of mineral-derived base oils by such of naturally regrowing and therefore rapidly biodegradable origin has been attempted in many cases. This approach seems to be obvious when considering the composition of greases with the base oil taking a share of up to 95% depending on the targeted grease consistency and application field. With regard to conformity to the demands of ecolabels such as the German Blue Angel, however, the mere substitution of the base oil with a rapidly biodegradable one is still insufficient. Also for the thickening agent, which is regarded a basic substance, biodegradability must be verified by at least 70% by using one of the test methods according to OECD 301 B, F, D or C, or ISO 14593 or ISO 10708.[4] Although biodegradable thickeners are scarcely found in current presence on the world market, previous investigations conducted in the field of biogenic thickening agents have been reported. Thus, the use of different biopolymers derived from natural constituents such as polysaccharides as thickening agents in the formulation of biologically and environmentally acceptable lubricating greases has been proposed.[4] In this sense, some formulations containing thickening agents derived from cellulose and chitin/chitosan and castor oil have been rheologically[5–7] and tribologically[8] characterised. The basic criteria for award of the Blue Angel environmental label also accept mineral thickeners in bio-greases. Taking into account these considerations, the main goal of this study is to investigate the tribological characteristics of mineral thickeners such as bentonite (BT) and highly dispersed silica acid (HDS). These were used as substitute thickeners for metal soaps to formulate rapidly biodegradable vegetable oils or fatty acid esters into greases. Although these mineral clay thickeners are usually used in tribocontacts where vegetable oils would fail (i.e. high temperature applications), they still qualify as biologically and environmentally inert thickeners. Although compliance verification has not been furnished officially, the formulated bio-greases investigated in this study theoretically fully comply with the aforementioned criteria of the Blue Angel. This ecolabel was used only exemplarily for many other labels with similar requirements; other comparable ecolabels are listed in Bartz[9] and discussed along with recommendation to their test methods in Battersby.[10]

MATERIALS AND METHODS

The Greases

Bentonite and HDS were used as thickening agents to formulate biodegradable greases by using low viscous polyalpha olefin (PAO), as non-biogenic reference oil, and three biogenic and rapidly biodegradable oils, high oleic sunflower oil (HOSO), octyldodecyl isostearate (OCT) and trimethylolpropane trioleate (TMPO). The latter three of which mainly consisted of esters of organic fatty acids. The main physicochemical base oil characteristics are collected in Table I.

The National Lubricating Grease Institute consistency grade 2, according to DIN 51818 and ISO 2137, was targeted for all greases. This resulted in unique amounts of thickener due to the different base oil viscosities. Compositions and physical grease characteristics are depicted in Table II.

Table I. Physicochemical base oil characteristics.

	v_{40} (mm^2 s^{-1})	v_{100} (mm^2 s^{-1})	Density (g cm^{-3})	Degradability
PAO (synth. reference oil)	46.7	7.9	0.83	Yes[a]
HOSO	38.8	8.5	0.92	96%[b]
TMPO	48.0	9.8	0.92	96%[b]
OCT	25.5	5.5	0.87	97%[b]

[a]According to Materials Safety Data Sheet, it is expected to be inherently biodegradable.
[b]According to OECD 301 B.
PAO: polyalpha olefin.
HOSO: high oleic sunflower oil.
TMPO: trimethylolpropane trioleate.
OCT: octyldodecyl isostearate.
synth: synthetic.

Table II. Grease composition and some physical grease characteristics.

| | Thickener | | Swelling | Pu[a] | Pw[a] | NLGI |
Grease label	Type	Content (%)	Aid (%)	(1/10 mm)	(1/10 mm)	Class
BT–PAO	Bentonite	20.1	1.07	249	279	2
BT–HOSO		28.5	1.12	257	268	2
BT–OCT		15.8	0.82	264	291	2
BT–TMPO		24.0	1.18	242	279	2
HDS–PAO	Highly dispersed silica acid	10.3	1.09	269	280	2
HDS–HOSO		13.9	1.43	238	268	2
HDS–OCT		11.4	0.89	264	287	2
HDS–TMPO		14.1	2.73	249	287	2

[a]Worked and unworked penetration according to DIN ISO 2137.
BT: bentonite.
PAO: polyalpha olefin.
HOSO: high oleic sunflower oil.
OCT: octyldodecyl isostearate.
TMPO: trimethylolpropane trioleate.
HDS: highly dispersed silica acid.
NLGI: National Lubricating Grease Institute.

Highly dispersed silica acid thickener consists of pyrogenic silica acid generated in flame hydrolysis. This process results in SiO$_2$ primary particles, which melt together and shape larger units of 0.1–1 μm of size in the flame reaction. When cooling down after the flame hydrolysis, these larger units link together in flaky chain agglomerates with very high specific surface (up to 400 m^2 g^{-1}). HDS is organophilised by cleavage of silanol groups through organosilicon compounds.[11] The thickening functionality of the agglomerates is based on outstanding hydroxyl groups, which enable hydrogen bridge linkages to the surrounding base oil and adjoining elements.

Bentonite thickener is obtained from impure clay of volcanic ashes. The main constituent of BT is montmorillonite, and also, other minerals such as quartz, mica, feldspar, pyrite and lime are found. All of which make up a chemical composition of up to 60% of SiO$_2$ and 20% of Al$_2$O$_3$ besides several metallic oxides.[11] The swelling and gel-forming abilities of BTs are based on the triple layer particle structure of montmorillonite along with its high ion exchangeability. Organophillic modification, primarily with quarternary ammonium salts, activates the surface of the layer particles with hydrocarbon molecules and thus gives them oleophilic properties.[11–14] In the processing of BT thickener, the addition of an activator (carbonates, glycols, ketones, alcohols and water) is needed first to separate agglomerates of layer particles along with mechanical shearing and, second, to enable these particles to build up hydrogen bridge linkages to each other.[3,11] Thus, the mineral clays form a three-dimensional network, which physically binds the base oil.

Tribological Tests

All formulated greases were tested using a ball-on-disc nanotribometer in rotational mode with sliding contact situation. The tribological contact was situated with a fixed ball (diameter of 1.5 mm) that was pressed to a revolving steel disc (115CrV3, hardness 22.72 HRC) with defined normal loads of 500, 100, 10 and 1 mN. This same experimental set-up was used in previously described studies. A more detailed description of the experimental set-up can be found in Fiedler.[15] Material combinations in the tests were varied using sapphire (99.9% polycrystalline Al_2O_3) and steel (100Cr6) balls. The relative velocity between the ball and disc was set at 5 mm s^{-1} for all tests. The greases to be examined were applied to the disc surface prior to the start up of each test offering the tribological contact a grease layer of 0.05 mm. The test duration was set at 50 min without manual relubrication of the steel plates. Because this fact may involve a danger of critical grease deficiency in the contact, the development of friction coefficient over time was observed thoroughly to ensure that no starvation effects occurred. In this way, it is guaranteed that tribocontacts were always fully flooded and operated under steady state condition.

RESULTS

Friction

Frictional results of all performed experiments are depicted in Figure 1 in logarithmic scale.

This bar diagram clearly shows that the friction coefficient directly depends on the applied normal load with a general tendency towards much higher values for low normal loads. In fact, values increase up to 10-fold when comparing the results of 1 mN with those achieved with 500 mN. An explanation for this effect might be found in the set-up of the experiment. All greases were applied to the steel discs only prior to the start of the test, resulting in the fact that the static test ball had to plough its way through the grease layer in every revolution of the plate. This ploughing offers a rheological resistance to the relative motion, which, although steady for all grease layers of comparable consistency and thickness, is insignificantly small at high normal loads but the main resisting factor for very small normal forces. Besides this, the group of BT greases provided lower friction coefficient values compared with HDS-thickened greases, independently of material combination and normal load. The only exception to this rule is BT–TMPO within the normal load regime of 1 mN — in both material combinations, it resulted in higher coefficients of friction compared with HDS–TMPO. Because of the very low normal load of 1 mN and its resulting rheological significance, explanations for this exception are presumed to be found rheologically. For this reason, further rheological grease investigation is suggested. Also, in both material combinations, the use of different base oils did not significantly influence the frictional results in the group of BT greases as much as it did within the group of HDS greases. In this group, HDS–PAO-based and HDS–OCT-based greases generally led

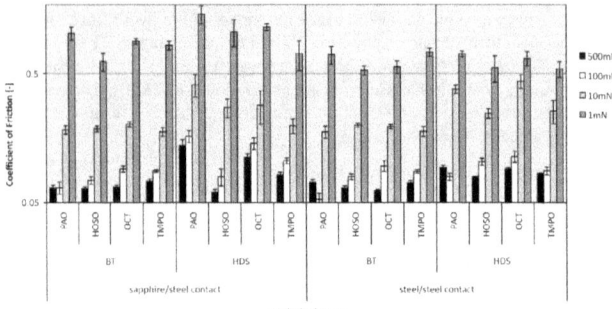

Figure 1. Friction of the tribosystems — both material combinations. BT, bentonite; HDS, highly dispersed silica acid; PAO, polyalpha olefin; HOSO, high oleic sunflower oil; OCT, octyldodecyl isostearate; TMPO, trimethylolpropane trioleate.

to higher values of the friction coefficient than HDS–HOSO-based and HDS–TMPO-based greases, for all applied normal loads. This difference is even more intensified with a change to a sapphire/steel material combination, generally resulting in higher frictional values at the application of HDS greases. The examined BT greases, however, showed a much more stable frictional response throughout both material couplings and no clear influences of both the thickeners and the base oils. Reasons for this behaviour are found in the micro-structural and physicochemical properties of the thickeners in combination with the base oils. These will be discussed later on after analysis of the wear behaviour of the described tribocontacts.

Wear

Figure 2 displays the widths of all wear marks in balls and discs, respectively, represented in bars. For a better overview and comparability, the values of the wear tracks in the discs were negated and their bars coloured in the same greyscale as those of the corresponding wear marks in the balls.

This diagram clearly shows several effects that can be distinctively associated with material combination, formulated thickener and applied base oil or a mixture of either one of them.

The analysis of wear results from the material combination point of view reveals different system responses depending on the thickener. This diagram clearly shows that the group of BT greases in sapphire/steel combination resulted in much less ball wear than did the HDS greases. Wear widths of HDS greases in some parts even amount to twice the value of those achieved with BT greases for the higher applied normal forces (500 and 100 mN). With lower normal loads (10 and 1 mN), however, wear marks created with HDS greases are only around 1.4-fold to 1.5-fold wider than the ones produced with BT greases. Besides this, whereas in sapphire/steel contact, BT greases resulted in the narrowest wear marks in the balls, in the whole series of experiments with an average value of 37.81 ± 9.51 μm, they produced very wide ball wear scars (78.44 ± 7.95 μm) in steel/steel contact. Most interestingly, this effect is slightly reversed for the group of HDS greases with very wide wear marks in the sapphire balls (63.38 ± 23.41 μm) but relatively narrow ones in the steel balls (60.00 ± 17.05).

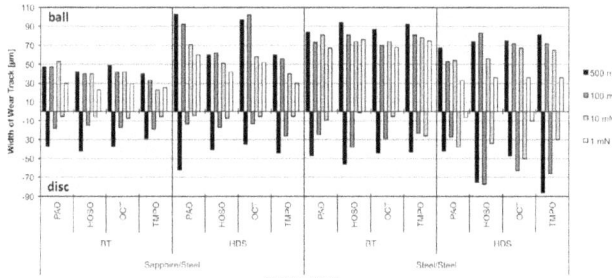

Figure 2. Widths of wear marks on balls and discs — both material combinations. Values of disc wear were negated for better overview and comparability. BT, bentonite; HDS, highly dispersed silica acid; PAO, polyalpha olefin; HOSO, high oleic sunflower oil; OCT, octyldodecyl isostearate; TMPO, trimethylolpropane trioleate.

Another conspicuousness that becomes apparent when assessing the results presented in Figure 2 from the material's point of view relates to the interaction of material combination and the base oils. On the one hand, in all sapphire/steel contacts, PAO and OCT esters in combination with both thickener types resulted in the widest wear marks produced on sapphire balls within their respective thickener group. In the group of HDS greases, the impact of this effect is most severe producing wear marks 57.1% wider than the ones resulting from the use of HOSO and TMPO esters. Although not as severe, this effect is still obviously apparent within the group of BT greases resulting in 28.2% wider wear marks (given numbers were averaged for all normal forces within the respective thickener group and depend on and generally diminish along with the applied normal load). On the other hand, the influence of the base oil on the widths of all ball wear marks is reversed in the steel ball on steel disc material combination. Here, HOSO and TMPO esters resulted in the widest wear marks (8.0% wider within the group of BT greases and 9.4% wider in the group of HDS greases) compared with those obtained with PAO and OCT esters.

The assessment of the present data also makes clear that whereas the wear marks in the specimen balls revealed the previously described clear distinction between BT and HDS greases or the different base oils, the widths of wear marks in the discs do not all show this same clear response. In fact, the wear widths in the discs of the greases tested in sapphire/steel contact resemble quite much; the only difference is that the wear widths produced with HDS greases are 14.8% wider in average. When comparing the extent of wear marks in the discs with the ones left in the sapphire balls, one finds no correlation of these two within the group of greases tested in sapphire/steel contact. The situation is different in steel/steel contact, however. Here, the same interactions predominate between formulated base oils and their respective wear widths as were found in the steel balls. Although they are only to be detected as general trends and for higher normal forces within the group of BT greases, they are clearly present for all HDS greases resulting in the widest wear marks when the contact is lubricated with HOSO-based and TMPO-based greases.

Micrographic appraisal of wear results in the retrospect of tribological tests gives direct indication on wear behaviour, which consists of prevailing wear mechanisms as well as wear types as defined by Fleischer or Czichos.[16,17] Figures 3–6 display the micrographs of wear results for some selected

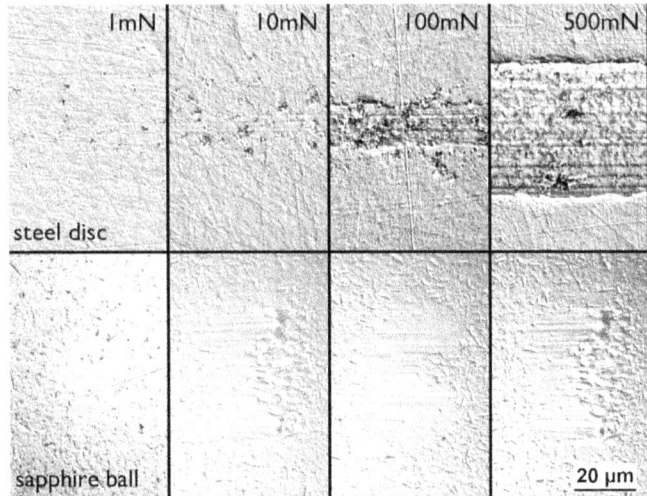

Figure 3. Bentonite–high oleic sunflower oil in material combination sapphire ball on steel disc.

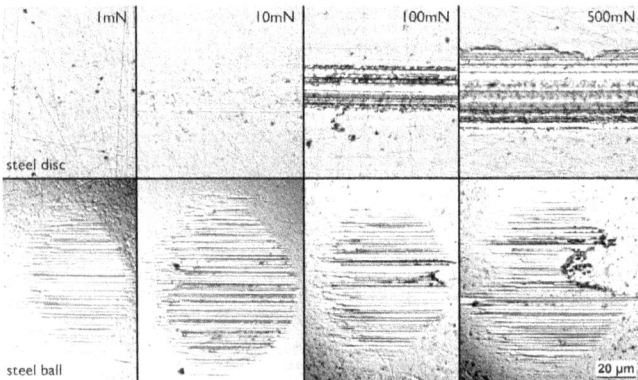

Figure 4. Bentonite–trimethylolpropane trioleate in material combination steel ball on steel disc.

tribosystems, which were assessed in the course of the experiments performed in this study. Each individual figure shows the wear marks of a specific tribosystem in both the disc and the ball examined

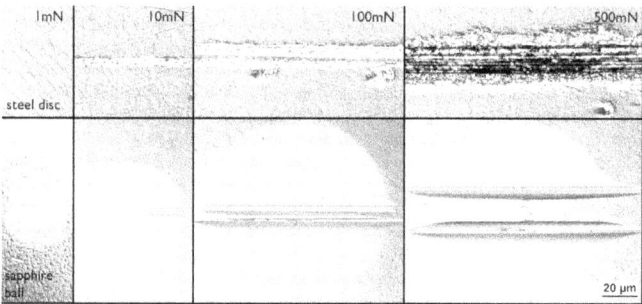

Figure 5. Highly dispersed silica acid–octyldodecyl isostearate in material combination sapphire ball on steel disc.

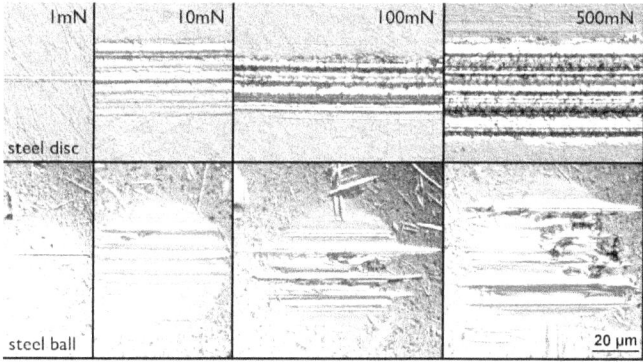

Figure 6. Highly dispersed silica acid–polyalpha olefin in material combination steel ball on steel disc.

at all given normal load regimes (1–500 mN). The influence of thickener type on wear response is clearly elucidated by these images. The use of the different base oils merely intensifies or reduces the thickener influences on the wear response, as previously described.

Bentonite Greases in Sapphire/Steel Contact. Figure 3 displays the wear responses of a tribosystem consisting of a sapphire ball on steel disc material combination containing BT–HOSO grease as lubricant. Focusing on the wear reaction of the steel discs reveals different types of wear mechanisms, each one predominating with different intensities under different normal load conditions. On the one hand, there is a wear track of uninterrupted plastic deformation on the discs in the central region of the ball contact with some signs of surface delamination. This wear mark increases in width along with the normal load applied and is not found for 1 mN. On the other hand, there is a band of speckles surrounding the just-described uninterrupted track of plastic deformation. The speckle size also increases with the normal load and completely combines with the plastic deformation in the case of 500 mN.

Focusing on the wear response of the sapphire balls mainly discloses two different wear mechanisms. The images of the sapphire balls tested in all applied normal loads show abrasive wear. The sapphire balls with applied normal loads 10, 100 and 500 mN even show a typical bean shape, which is repeatedly observed exclusively with the use of BT greases. In some parts, this characteristic shape is accompanied by additional adhesive components. This bean shape consists of two deeper grooves of most intensive abrasive wear of a depth of up to 166 nm and a distance of up to 24 μm leaving the main contact region in the centre less affected. Moreover, the sapphire balls of the three just-mentioned normal loads show quite similar wear intensities. This leads to the presumption of a critical load of 10 mN for the occurrence of this mechanism in the BT grease-lubricated contact. The aforementioned investigation of the width of wear tracks seems to corroborate this theory because all wear widths in sapphire balls created by applying 10 mN and more, in combination with BT greases, show similar values.

Bentonite Greases in Steel/Steel Contact. Figure 4 displays the wear results of a tribosystem lubricated with BT–TMPO in a steel ball on steel disc material combination. Analysis of the wear marks produced in the steel plates also discloses several wear mechanisms. First, wear marks of mainly abrasive nature in the central contacting area of the steel plate surface can be detected. These marks are most clearly visible in normal load regimes of 500 and 100 mN. Whereas the image corresponding to 10 mN still reveals slight abrasive lines, there are no such wear tracks by applying 1 mN normal load. Second and just like in sapphire/steel contact, there is a band of evenly distributed speckles to be detected around the central contacting area.

The circular wear marks in the steel balls also reveal a highly abrasive wear mechanism with wear grooves of an average width of 1.26 μm for all normal loads. The characteristically waisted wear mark shape produced in a normal load regime of 500 mN is found for all BT greases tested.

Highly Dispersed Silica Acid Greases in Sapphire/Steel Contact. The wear marks produced on the discs in a sapphire/steel contact situation lubricated with HDS greases also show a highly abrasive character in the central contacting area, as illustrated in Figure 5. Wear tracks also disclose a high proportion of plastic deformation when applying high normal forces (500 and 100 mN), resulting in a deep groove with high flanks. The application of 1 mN normal load results in no detectable wear at all.

The sapphire balls as counterparts in this contact also show a high rate of abrasive material removal from the surface, visible as a circular worn off cap. Flanks, which showed as elevations in the discs, appear as corresponding grooves in the sapphire balls. More interesting, however, is the fact that the rest of the circular worn off area appears as a smooth, almost evenly polished region, although the corresponding counterpart of the steel disc contact appears completely unspoiled. This circumstance seems even more abnormal when considering the hardness of sapphire compared with the hardness of the steel discs. This effect has been found in previous studies, and explanations for the same were provided in other publications.[15] An ensuing atomic force microscopy (AFM) analysis will elucidate mechanisms that favour this effect.

Highly Dispersed Silica Acid Greases in Steel/Steel Contact. The highly abrasive character of HDS-thickened greases in steel/steel contact is depicted very clearly in Figure 6. Here also, abrasion appears to be the main wear mechanism in both steel plates and sapphire balls. The aforementioned property of

HDS greases resulting in polishing of the surface of the balls while leaving contacted parts of the discs unaffected is also observed in steel/steel contact, especially when applying 1 mN.

DISCUSSION

Creation of 'Speckle Patterns' around the Main Contacting Area with the Use of Bentonite Greases

The AFM investigations of the worn surfaces provide more information about the nature of these very typical speckles as shown in Figure 7. They seem to consist of elevations with very nearby wholes in dimensions depending on the age of the speckle.

Figure 7 shows two line profiles through a main deformational wear track (3) with one adjacent speckle each, (1) and (2). The one that is referred to as point of emphasis number (1) presents itself as a whole of 74 nm in depth with an adjoining elevation of 80 nm in height. The speckle that is flagged in point number (2) is exclusively structured as an elevation (height 84 nm). The formation process of these speckles is assumed to commence at any time in the course of the experiment with the interlinking of small montmorillonite particles of the BT thickener to the surface of the steel plates. The physically working adhesive mechanisms will be discussed later. The fact that the corresponding sapphire counter surface wears off in most parts of the regions where these speckles occur indicates to extreme hardness of the speckles. This presumption becomes even more profound regarding the fact that the steel surface around the speckles appears completely unspoiled, although the sapphire is much harder than the steel. In further progression of this process, it seems plausible that other particles are more likely to adhere to these protruding elements and thus allowing the speckle to grow. Therefore, it is deduced that these protruding particles offer a target to the friction process resulting in high shearing stress in their base,

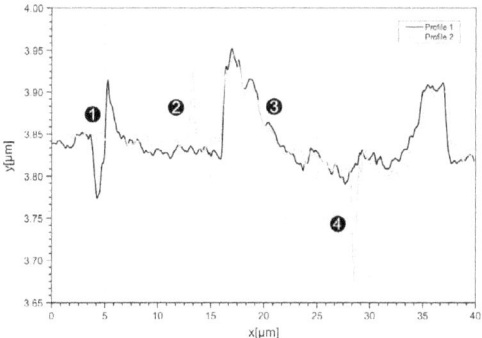

Figure 7. Atomic force microscopy profile of speckles and wear track — bentonite grease 100 mN in material combination sapphire ball on steel disc. Points of emphasis numbers 1–4: (1) older speckle in distance of 22.5 μm from the centre of the main groove (depth 74 nm, height 80 nm); (2) younger speckle in distance of 13.9 μm from the centre of the main groove (height 84 nm); (3) flank of main groove (height 108 nm, total penetration depth 52 nm); (4) older speckle in the centre or the main groove (depth 176 nm).

which will presumably plastically deform subjacent regions resulting in very nearby wholes. Therefore, it is concluded that since there is no adjoining whole in the speckle identified in point of emphasis number (2) of Figure 7, it is younger than the one marked as number (1). Another older speckle is indicated with point (4). It is located in the very centre of the main deformational groove, and by its depth (176 nm), it is presumed to be more advanced than the ones described so far. Also, it is assumed that it was produced through the previously described process apart from the fact that the typical elevation was cut off as a result of higher surface pressure in the main contacting area between the ball and disc in the course of the experiment. An indication of this theory is the appearance of the wear mark that evolved with a normal load of 500 mN in the steel disc as depicted in Figure 3 — here, it shows very clearly that all speckles were consumed by the deformational wear mark.

Highly Abrasive Wear Behaviour of Bentonite and Highly Dispersed Silica Acid Thickeners

The highly abrasive character of all examined mineral thickeners might be explained by their microphysical constitution. All of the previously presented images of wear appearance indicate that the microstructure of BT and HDS greases differ significantly. With the application of HDS greases, there were typically polished regions found in the specimen balls in both material combinations, whereas BT greases generally led to rougher worn off caps in the balls. Figure 8 shows AFM images of two selected greases containing these two thickeners.

The AFM analysis was performed in tapping mode with equal conditions and experimental set-up as reported in Sanchez.[18] The displayed AFM image of BT–HOSO grease reveals a large amount of evenly distributed solid platelets of montmorillonite particles easily spotted in white to bright grey scale. These particles are surrounded by the liquid base oil, which appears as dark grey to black regions. The AFM image of HDS–OCT also delineates solid particles, but in this case, SiO_2 particles are much smaller and appear distributed in the form of clusters. It is presumed that these smaller and harder particles induce a polishing material removal process in the contacting specimen balls. These results are in agreement with other studies previously reported,[19] where the effect of the tribologically

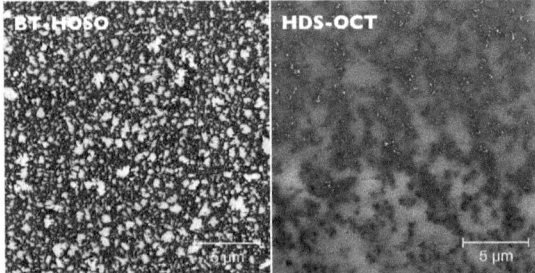

Figure 8. Atomic force microscopy micrographs of exemplary BT-thickened and HDS-thickened greases. BT–HOSO, bentonite–high oleic sunflower oil; HDS–OCT, highly dispersed silica acid–octyldodecyl isostearate.

active SiO_2 nanoparticles was investigated. A comparison of these results with the present study suggests SiO_2 particle sizes of less than 100 nm, as may be corroborated in Figure 8.

Base Oil-dependent, Thickener Type-dependent and Material Combination-dependent Frictional and Wear Responses

Reasons for this monitored behaviour are found in the molecular properties of the base oils, which mostly influence the propensity of oils to physically and chemically interact with surfaces. Further explanation of this subject will elucidate the effects experimentally found.

It is assumed that the polarity of the given base oils highly influences the interaction of the whole base-oil-thickener-surface system. Generally, the dimension and types of these interactions depend on the bonding character of the base oils and all interacting surfaces as described in Maßmann.[20] This bonding character is based upon secondary valence bond forces or van der Waals interactions of either Keesom, or Debye or London type depending on the existence and the dimension of dipole moments in both the oil and the solid surface. Due to these microphysical mechanisms in many cases, the affinity between oils and metal surfaces leads to the formation of separating films by adsorption processes. In current state of research, these surface-active properties are attributed to the polarity[21] and the extent of van der Waals interactions[20] between oils and surfaces. Many of the biodegradable oils' positive tribological characteristics are based on their highly polar molecule composition. This is why biodegradable oleochemicals especially esters of glycerine with long chain saturated and non-saturated fatty acids in many cases even qualify as lubricant additives[21] whose functionality is based on their polar-active character creating physically adsorbing, surface-separating films[22,23] in anti-wear, mild extreme-pressure, friction-modifying and rust-inhibiting additive packages.[23]

In the studies of Lugscheider[21] and Maßmann,[20] oil polarity is determined by means of measuring the polar shares of surface energy of the oils with the application of equations by either Wu,[24] or Owens and Wendt[25] or Fowkes.[26–28]

The current state of research suggests several ways to obtain the polarity of oils. Some authors make use of solubilisation methods,[29] and others use gas and other chromatographic methods such as high-performance liquid chromatography[30–33] and interfacial tension measurements between oil and water[34] to obtain oil polarities. El-Mahrab-Robert[35] compared some of these methods with a result of superiority of the two latter-mentioned ones.

Following the prior disquisition, the bonding types resulting from the given base oils and solid surfaces in this study have to be interpreted according to their specific polarities. In this study, the relative polarity of the given base oils is obtained by interfacial tension measurements between oil and water by using equivalent equipment and methods to the ones described in Ringard-Lefebvre[34] and El-Mahrab-Robert.[35] Results of these measurements are depicted in Figure 9. Interpretation of these results (the higher the interfacial tension, the lower is the oil polarity) leads to an order of relative polarities of the four investigated base oils, which can be described by OCT ≈ PAO ≪ HOSO ≈ TMPO starting from the oil of least relative polarity and ending with the highest one.

The polarities of the tribologically investigated surfaces in this study are taken from relevant literature[36] and can be ranked as Al_2O_3 < 115CrV3 ≈ 100Cr6. Yet, one must bear in mind that the prevailing main types of van der Waals interactions may change in the course of a test, especially taking into account the abrasion of highly polar oxide layers in metal surfaces. To elucidate the base oil polarities' influence on tribological interactions of the investigated surfaces, the aforementioned tests were repeated under equal conditions with pure base oils. The results depicted in Figure 10 very clearly indicate the presence

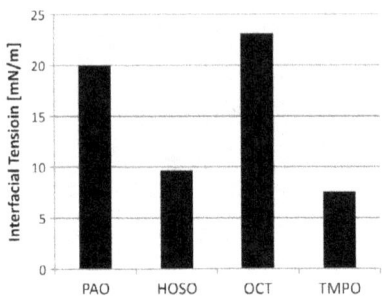

Figure 9. Interfacial tensions between distilled water and pure base oils. PAO, polyalpha olefin; HOSO, high oleic sunflower oil; OCT, octyldodecyl isostearate; TMPO, trimethylolpropane trioleate.

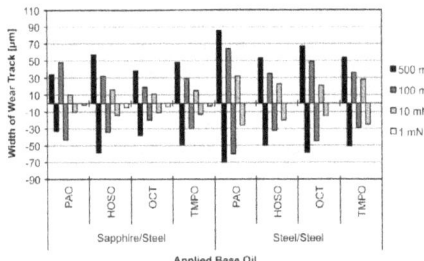

Figure 10. Widths of wear marks on balls and discs — both material combinations, pure base oil. Values of disc wear were negated for better overview and comparability. PAO, polyalpha olefin; HOSO, high oleic sunflower oil; OCT, octyldodecyl isostearate; TMPO, trimethylolpropane trioleate.

of a protective layer, which is formed due to the high polarity of HOSO and TMPO and its resulting Keesom bonds in physical adsorption processes.

These are made manifest by the comparably narrow wear scars with the application of the respective oils in steel/steel combination and in both contacting partners. On the other hand, the lower polarity oils PAO and OCT resulted in wider wear marks and thus offered less protection. This effect is reversed, however, in sapphire/steel contact — here, the high-polarity oils offered least protection, which might be ascribed to the smaller relative polarity of sapphire and to poorer wettability behaviour in the combination of sapphire solid surfaces and highly polar oils.

When applying these conclusions to the interpretation of tribological investigations of the given clay-thickened grease systems, one finds a perfect reversal of polarity influences in comparison of wear results of the pure oils (Figure 10) with those achieved with the respective greases (Figure 2). When considering the reasons for this effect, one must keep in mind which mechanisms may work within the greases and between the surfaces. Several, partially opposing, mechanisms come into effect with the use of high-polarity oils in combination with clay thickeners and differing material combinations.

For one thing, there is the influence of the thickening process: high-polarity oils, more than low-polarity ones, tend to deposit to outstanding silanol groups in primary particles of HDS greases and to hydroxyl groups of organosilicon particles in BT greases, thus interfering with and partially preventing the formation of a three-dimensional network and causing a higher need of share of thickener content for equal mechanical stability of the clay grease as shown in Table II. In this table, it is evident that more thickener is needed in combination with highly polar oils than with low-polarity oils. Moreover, the influence of this effect is bigger in BT grease systems than in HDS greases. This may be substantiated by the use of quarternary ammonium ions for organophillic modification of the BT greases, which were kept the same for all given BT greases in the process of formulation. This resulted in similar hydrocarbon chain rest molecules, which in turn deeply influenced the bonding character between thickeners and oils of different polarities. For further investigation of this subject, it is proposed to use tailor-made quarternary ammonium salts to fit the polarity of each base oil. This way, the share of thickener content would be kept at a fixed level, and the direct influences of base oil polarities could be investigated without interference of thickener content. Also for further investigation, in this context, it is proposed to examine the rheological grease behaviour depending on base oil polarity. In conclusion, this effect leads to the creation of a physisorbtive protective layer of highly polar oil surrounding the extremely hard silicon oxide compounds of both clay grease types. This may explain the comparably narrower wear tracks in the sapphire balls in combination with high-polarity greases BT–HOSO and BT–TMPO, even though more of these hard particles are present.

For another thing, there is the influence of physisorbtive adhesion processes in association with highly polar oils and metallic tribological surfaces, which leads to less wear with the use of pure oils as described earlier. In combination with clay thickeners, however, metal oxide layers of tribological surfaces seem to more attract highly polar oils than clay particles and thus leading to a destruction of the just-described protective layer surrounding them. This, however, leads to a higher number of unshielded extremely hard clay particles in the tribological steel/ steel contact resulting in higher ware rates for clay greases with highly polar oils than with low-polarity ones.

The aforementioned disquisition on mutual interdependence of bonding mechanisms between oils, solid body surfaces and thickener surfaces leaves room to conclude that the same mechanisms may lead to adhesion of clay thickener particles to tribological surfaces. This may elucidate the 'speckle'-forming process discussed in the Creation of 'Speckle Patterns' around the Main Contacting Area with the Use of Bentonite Greases Section of this paper. Moreover, it has been observed during the evaluation of wear images taken after the tests that highly polar oils generally tend to support the aforementioned speckle forming. This may add another opponent to the protective character of highly polar oils in clay-thickened greases: Provided that these speckles also form on the surface of the investigated steel balls, it may be deduced that they add to the highly abrasive character of BT greases when they come into contact with the speckles on the steel plates during the course of the experiment. Therefore, it is assumed that the speckles especially have a negative influence on the wear behaviour if they occur in both contacting surfaces.

CONCLUSIONS

The investigation of greases based on biodegradable base oils and mineral thickeners BT and HDS revealed a clear distinction between each main component's influence on frictional and wear behaviour. Findings moreover disclose a different behaviour of these components depending on the material combination. In sapphire/steel material combination, BT greases resulted in much less wear than did the HDS greases. However, BT greases showed wider wear marks in steel/steel contact. Base oil influence on wear behaviour is much more apparent in HDS-thickened than in BT-thickened grease systems, which is otherwise inverted with a change of material combination. Frictional and wear results correlate very well in sapphire/steel contact resulting in high coefficients of friction for tribocontacts with high wear rates. In steel/steel contact, however, this effect is not found.

The unique interactions of base oil polarities and the clay thickener particles with its resulting differences in share of content result in a perfect reversal of base oil polarity influences in wear responses to tribological tests.

REFERENCES

1. Willing A. Lubricants based on renewable resources – an environmentally compatible alternative to mineral oil products. *Chemosphere* 2001; **43**(1):89–98. DOI: 10.1016/S0045-6535(00)00328-3
2. Wagner H. Lubricant base fluids based on renewable raw materials: their catalytic manufacture and modification. *Applied Catalysis A: General* 2001; **221**(1–2):429–442. DOI: 10.1016/S0926-860X(01)00891-2
3. Dresel W. Biologically degradable lubricating greases based on industrial crops. *Industrial Crops and Products* 1994; **2**(4):281–288. DOI: 10.1016/0926-6690(94)90119-8
4. RAL gGmbH. Basic criteria for award of the environmental label – readily biodegradable lubricants and forming oils, *RAL-UZ 64* 2011.
5. Sanchez R. Thermal and mechanical characterization of cellulosic derivatives-based oleogels potentially applicable as bio-lubricating greases: influence of ethyl cellulose molecular weight. *Carbohydrate Polymers* 2011; **83**(1):151–58. DOI: 10.1016/j.carbpol.2010.07.033
6. Sanchez R. Use of chitin, chitosan and acylated derivatives as thickener agents of vegetable oils for bio-lubricant applications. *Carbohydrate Polymers* 2011; **85**(3): 705–714. DOI: 10.1016/j.carbpol.2011.03.049
7. Sanchez R. Rheological and mechanical properties of oleogels based on castor oil and cellulosic derivatives potentially applicable as bio-lubricating greases: influence of cellulosic derivatives concentration ratio. *Journal of Industrial and Engineering Chemistry* 2011; **17**(4):705–711. DOI: 10.1016/j.jiec.2011.05.019
8. Sanchez R. Tribological characterization of green lubricating greases formulated with castor oil and different biogenic thickener agents: a comparative experimental study. *Industrial Lubrication and Tribology* 2011; **63**(6):446–452. DOI: 10.1108/00368791111169034
9. Bartz W. Ecotribology: environmentally acceptable tribological practices. *Tribology International* 2006; **39**(8):728–733. DOI: 10.1016/j.triboint.2005.07.002
10. Battersby N. The biodegradability and microbial toxicity testing of lubricants – some recommendations. *Chemosphere* 2000; **41**(7):1011–1027. DOI: 10.1016/S0045-6535(99)00517-2
11. Goerz T. Gel-und bentonitfette zusammensetzung-eigenschaften. *Tribologie und Schmierungstechnik* 2009; **56**(2):30–34.
12. Pogosyan A. Tribological properties of bentonite thickener-containing greases. *Journal of Friction and Wear* 2008; **29**(3): 205–209. DOI: 10.3103/S1068366608030094
13. Chtourou M. Modified smectitic Tunisian clays used in the formulation of high performance lubricating greases. *Applied Clay Science* 2006; **32**(3–4):210–216. DOI: 10.1016/j.clay.2006.03.003
14. Kohashi H. Application of fatty acid esters for lubricating oil, in *World Conference on Oleochemicals into the 21st Century*, Applewhite T (Ed.) American Oil Chemists' Society, Champaign, Illinois, 1991: 243–247.
15. Fiedler M. Tribological properties of greases based on biogenic base oils and traditional thickeners in sapphire-steel contact. *Tribology Letters* 2011; **44**(3):293–304. DOI: 10.1007/s11249-011-9848-9
16. Fleischer G. *Verschleiß und Zuverlässigkeit*. VEB Verlag Technik, Berlin, 1980.
17. Czichos H. *Tribologie-Handbuch*. Vieweg und Teubner, Wiesbaden, 2010.

18. Sanchez M. Atomic force microscopy and thermo-rheological characterisation of lubricating greases. *Tribology Letters* 2011; **41**(2):463–470. DOI: 10.1007/s11249-010-9734-x
19. Peng D. Size effects of SiO_2 nanoparticles as oil additives on tribology of lubricant. *Industrial Lubrication and Tribology* 2010; **62**(2):111–120. DOI: 10.1108/00368791011025656
20. Maßmann T. *Wirkmechanismen Additivierter Schmierstoffe in der Kaltumformung*. Shaker, Aachen, 2007.
21. Lugscheider E. Wettability of PVD compound materials by lubricants. *Surface and Coatings Technology* 2003; **165**(1):51–57. DOI: 10.1016/S0257-8972(02)00724-7
22. Korff J. *Additive für Schmierstoffe*. Expert Verlag, Renningen-Malmsheim, 1994: 280–301.
23. Meyer K. *Additive für Schmierstoffe*. Expert Verlag, Renningen-Malmsheim, 1994: 392–418.
24. Wu S. Calculation of interfacial tension in polymer systems. *Journal of Polymer Science Part C: Polymer Symposia* 1971; **34**(1):19–30. DOI: 10.1002/polc.5070340105
25. Owens D, Wendt R. Estimation of the surface free energy of polymers. *Journal of Applied Polymer Science* 1969; **13**(8):1741–1747. DOI: 10.1002/app.1969.070130815
26. Fowkes F. Additivity of intermolecular forces at interfaces. I. Determination of the contribution to surface and interfacial tensions of dispersion forces in various liquids. *The Journal of Physical Chemistry* 1963; **67**(12):2538–2541. DOI: 10.1021/j100806a008
27. Fowkes F. Attractive forces at interfaces. *Industrial and Engineering Chemistry* 1964; **56**(12):40–52. DOI: 10.1021/ie50660a008
28. Fowkes F. Donor-acceptor interactions at interfaces. *The Journal of Adhesion* 1972; **4**(2):155–159. DOI: 10.1080/00218467208072219
29. Isengard H. Determination of total polar material in frying oil using accelerated solvent extraction. *Lipid Technology* 2010; **22**(6):134–136. DOI: 10.1002/lite.201000019
30. Dobarganes M. Determination of polar compounds, polymerized and oxidized triacylglycerols, and diacylglycerols in oils and fats. *Pure and Applied Chemistry* 2000; **72**(8):1563–1575.
31. Knorn B. Bestimmung der polarität gesättigter monoglyceride durch gaschromatographie. *Food/Nahrung* 1977; **21**(9):817–823. DOI: 10.1002/food.19770210911
32. Rouser G. Determination of polar lipids: quantitative column and thin-layer chromatography. *Journal of the American Oil Chemists' Society* 1965; **42**(3):215–227. DOI: 10.1007/BF02541135
33. Márquez-Ruiz G. Rapid, quantitative determination of polar compounds in fats and oils by solid-phase extraction and size-exclusion chromatography using monostearin as internal standard. *Journal of Chromatography. A* 1996; **749**(1–2):55–60. DOI: 10.1016/0021-9673(96)00429-3
34. Ringard-Lefebvre C. Effect of spread amphiphilic [beta]-cyclodextrins on interfacial properties of the oil/water system. *Colloids and Surfaces. B, Biointerfaces* 2002; **25**(2):109–117. DOI: 10.1016/S0927-7765(01)00297-1
35. El-Mahrab-Robert M. Assessment of oil polarity: comparison of evaluation methods. *International Journal of Pharmaceutics* 2008; **348**(1–2):89–94. DOI: 10.1016/j.ijpharm.2007.07.027.
36. Becker G. *Kunststoff-Handbuch: Duroplaste*. Hanser, München, 1988.

www.ingramcontent.com/pod-product-compliance
Lightning Source LLC
Chambersburg PA
CBHW071757200526
45167CB00017B/368